Principles of Computational Genomics

The advent of high-throughput experimental assays, and in particular of next-generation sequencing, has revolutionized life sciences by enabling the generation of data at the scale of the whole genome. Extracting biologically useful or clinically actionable information from this data requires analytical methods quite different from the ones used to analyze low-throughput experimental results. The development of these methods is the goal of computational biology.

Understanding the principles on which these methods are based is thus necessary for all students and researchers in life sciences. This book provides the conceptual framework needed to understand computational genomics enough to be able to follow the arguments in recent papers, or to collaborate with computational scientists in research projects. In particular, it introduces the mathematical and statistical basis of the methods in some depth. The main focus is on the analysis of next-generation-sequencing assays, both for the interpretation of the DNA sequence per se (sequence alignment, phylogenetic tree reconstruction, genetic variants) and for the study of gene regulation and epigenomics (gene expression, transcription factor binding, chromatin states, 3D structure of the genome). The final chapter discusses the associations of genetic variants with phenotypes and diseases, and their biological interpretation.

Principles of Computational Genomics provides a solid foundation for understanding the many parts of computational genomics, including those not treated directly in the book. It will be of great benefit to students and researchers across the life sciences.

Chapman & Hall/CRC
Computational Biology Series

About the Series:

This series aims to capture new developments in computational biology, as well as high-quality work summarizing or contributing to more established topics. Publishing a broad range of reference works, textbooks, and handbooks, the series is designed to appeal to students, researchers, and professionals in all areas of computational biology, including genomics, proteomics, and cancer computational biology, as well as interdisciplinary researchers involved in associated fields, such as bioinformatics and systems biology.

Systems Biology and Machine Learning Methods in Reproductive Health
Abhishek Sengupta, Priyanka Narad, Dinesh Gupta and Deepak Modi

Computational Intelligence for Oncology and Neurological Disorders
First Edition
Mrutyunjaya Panda, Ajith Abraham, Dr. Gopi Biju and Dr. Reuel Ajith

Stochastic Modelling for Systems Biology
Third Edition
Darren J. Wilkinson

Computational Genomics with R
Altuna Akalin, Bora Uyar, Vedran Franke, Jonathan Ronen

An Introduction to Computational Systems Biology: Systems-level Modelling of Cellular Networks
Karthik Raman

Virus Bioinformatics
Dmitrij Frishman, Manuela Marz

Multivariate Data Integration Using R: Methods and Applications with the mixOmics Package
Kim-Anh LeCao, Zoe Marie Welham

Bioinformatics
A Practical Guide to NCBI Databases and Sequence Alignments
Hamid D. Ismail

Data Integration, Manipulation and Visualization of Phylogenetic Trees
Guangchuang Yu

Bioinformatics Methods
From Omics to Next Generation Sequencing
Shili Lin, Denise Scholtens and Sujay Datta

Systems Medicine
Physiological Circuits and the Dynamics of Disease
Uri Alon

The Applied Genomic Epidemiology Handbook
A Practical Guide to Leveraging Pathogen Genomic Data in Public Health
Allison Black and Gytis Dudas

Principles of Computational Genomics
Paolo Provero

For more information about this series please visit:
https://www.routledge.com/Chapman--HallCRC-Computational-Biology-Series/book-series/CRCCBS

Principles of Computational Genomics

Paolo Provero

CRC Press
Taylor & Francis Group
Boca Raton London New York

CRC Press is an imprint of the
Taylor & Francis Group, an **informa** business

A CHAPMAN & HALL BOOK

First edition published 2025
by CRC Press
2385 NW Executive Center Drive, Suite 320, Boca Raton FL 33431

and by CRC Press
4 Park Square, Milton Park, Abingdon, Oxon, OX14 4RN

CRC Press is an imprint of Taylor & Francis Group, LLC

ISBN: 978-1-032-58400-3 (hbk)
ISBN: 978-1-032-58399-0 (pbk)
ISBN: 978-1-003-44992-8 (ebk)

DOI: 10.1201/9781003449928

Typeset in Latin Modern font
by KnowledgeWorks Global Ltd.

To Elizabeth, Emma, and Martin

Contents

Preface

The advent of high-throughput experimental assays, and in particular of next-generation sequencing, has revolutionized life sciences by enabling the generation of data at the scale of the whole genome. Extracting biologically useful or clinically actionable information from this data requires analytical methods quite different from the ones used to analyze low-throughput experimental results. The development of these methods is the goal of computational biology.

Understanding the principles on which these methods are based is thus necessary for all students and researchers in life sciences. This does not mean that all need to learn how to *perform* such analyses, since most projects involving high-throughput data are carried out by collaborations between experimental and computational scientists. This book aims to provide students and researchers with the conceptual framework needed to understand computational genomics enough to be able to follow the arguments in a recent paper, or to collaborate with computational scientist in a research project. In particular, we discuss the mathematical and statistical basis of the methods in some depth, but not their implementation in terms of computer code.

In my experience, it is easier (at least for most people) to understand a statistical method when it is presented in the context of a concrete scientific question to be answered starting from some data. Thus, each method is developed starting from a biological question and a real (or, in a few cases, synthetic) dataset.

Chapter 1 is devoted to the comparison of biological sequences. For example, we tackle the following problem: Given a set of biological sequences that we observe in species living today, reconstruct the evolutionary history by which they derived from a common ancestor. The mathematical tools we introduce are *sequence alignment* and the *reconstruction of phylogenetic trees*, which requires introducing some concepts of *graph theory*. Moreover, we introduce the basic concepts of *comparative genomics*. We also discuss in this chapter the alignment to the reference genome of the short reads produced by next generation sequencing techniques.

In Chapter 2, we switch to transcriptomics, starting with the most basic question: Find the genes that have different expression levels when comparing two or more biological conditions. This leads us to introduce some of the main statistical ideas used in computational genomics (and in all other data-intensive disciplines), namely *hypothesis testing*, *regression*, and *likelihood*.

Chapter 3 focuses on how to use transcriptomics to understand gene function through the concept of guilt by association: Sets of genes characterized by similar expression patterns across biological conditions are identified through *clustering*, and their common function through *enrichment analysis*. We then show that the same methods can be used to cluster samples, rather than genes, and we show how this idea, applied to cancer transcriptomics, allows the identification of tumor molecular subtypes which can have clinical relevance, as demonstrated by *survival analysis.*

Chapter 4 concludes our treatment of transcriptomics by looking at single-cell RNA-seq data. *Dimensional reduction* is used to visualize the data in a way that is interpretable by our brains, and to facilitate data analysis. We introduce some more elements of graph theory, which provide the basis for new methods of class discovery and for *trajectory inference.*

The focus of Chapter 5 is gene regulation, the mechanism behind the patterns of gene expression explored in the previous three chapters. Starting from ChIP-seq data, we learn how to identify the genomic regions bound by a transcription factor using *peak calling*, while *motif analysis*, based on positional frequency matrices, allows us to identify the specific DNA sequence recognized by the transcription factor. Other sequence features, such as CpG islands, cannot be analyzed using positional matrices, which leads us to the introduction of *Markovian models.* These are also used to integrate different ChIP-seq signals related to epigenetic marks so as to obtain a segmentation of the genome into chromatin states. Finally, we briefly discuss how HiC data are analyzed to reveal the three-dimensional conformation of the genome, and its relevance to gene regulation.

Chapter 6 deals with the analysis of genetic variants, that is, the differences in DNA sequence between individuals of the same species. Here, the biological problem is to understand whether and how a given variant affects the phenotype. Regarding coding variants, we show how the combination of concepts from genetics and molecular biology can be used to prioritize the variants found in an individual affected by a Mendelian disease, so as to pinpoint the ones most likely to have caused the disease. Then we discuss complex traits, and specifically the tools that can help us identify causal variants among those associated to a trait by genome-wide association studies. *Machine learning* is introduced in this context, using as an example the problem of predicting the effects of a genetic variant on the state of the chromatin.

The book assumes that the reader is acquainted with modern molecular biology and genetics at the undergraduate level, and with mathematics (including probability and statistics) at the level of introductory undergraduate courses. As already mentioned, we do not discuss computer programming (except for mentioning a few useful software packages): Therefore, this book will not provide the reader with the tools needed to become a computational biologist. Many excellent books are devoted to this subject. However, I believe that the concepts discussed here will be useful also to aspiring computational biologists in understanding the rationale behind the software tools they use to analyze data.

This is definitely not a reference book, as many important topics are just mentioned in passing or skipped altogether. The choice of topics and examples is strongly based on my personal knowledge and taste. However, I believe that the concepts developed here can provide a solid foundation for understanding also the many parts of computational genomics that are not treated in the book.

Finally, I am grateful to many hundreds of students at the University of Turin who endured my undergraduate and graduate courses on computational biology, and provided plenty of feedback both directly and indirectly. I am especially grateful to my friends and longtime collaborators Davide Cittaro and Davide Marnetto who read successive versions of this book and provided many suggestions, all of them insightful and some that I actually accepted. It goes without saying that all remaining errors are my own.

Author Bio

Paolo Provero obtained his PhD in Theoretical Physics from the University of Genoa in 1992. Since 2000, his research focuses on computational genomics and, in particular, on the evolution and variation of gene expression and gene regulation. He is currently Full Professor of Molecular Biology at the University of Turin, Italy, where he teaches several courses in computational biology at the undergraduate and graduate level.

CHAPTER 1

Sequence Alignment and Phylogenetics

1.1 INTRODUCTION

Biological information is encoded in sequences (of nucleotides or amino acids), and sequence analysis and comparison are among the fundamental tasks of computational genomics. In this chapter, we describe the computational methods used to compare nucleotide or amino acid sequences, in particular, using *sequence alignment* to evaluate their similarity.

Sequences can be similar because they share a biological function and/or because of common descent. In the latter case, sequence alignment can be used as the basis to reconstruct the evolutionary history that led from a common ancestor to a set of extant sequences through the accumulation of mutations. We can infer such history using *phylogenetic tree reconstruction*, which requires the introduction of the simplest concepts of *graph theory*. When applied to individual genes, phylogenetic tree reconstruction allows us to define the concepts of *homology*, *orthology*, and *paralogy*. We will also see how sequence alignment provides the foundation of the analysis of next-generation sequencing data.

1.2 SEQUENCE ALIGNMENT

1.2.1 Pairwise Sequence Alignment

Suppose we are given two[1] biological sequences (DNA, RNA, or protein) that we want to compare. A *pairwise sequence alignment* is a way of displaying the sequences so as to highlight and assess their similarities and differences. Let us start with two very simple DNA sequences:

$$S_1 = GATTACA; \quad S_2 = GTTAAGA$$

---

[1] For technical reasons, the cases of two or more than two sequences are treated separately, and for the time being we consider only the comparison of two sequences.

DOI: 10.1201/9781003449928-1

> ℹ️ **Pairwise sequence alignment**
>
> A pairwise alignment of two sequences is a way of displaying them on top of each other, with the possibility of inserting the *indel* sign, "-", in any position of each sequence, as long as no column contains only indel signs.

The reason why the symbol "-" is called *indel* will become clear shortly. Thus,

$$
A_1 = \begin{array}{ccccccccc}
1 & 2 & 3 & 4 & 5 & 6 & 7 & 8 \\
G & A & T & T & - & A & C & A \\
G & - & T & T & A & A & G & A
\end{array}
$$

$$
A_2 = \begin{array}{ccccccccc}
1 & 2 & 3 & 4 & 5 & 6 & 7 & 8 & 9 \\
G & A & - & - & T & T & A & C & A \\
G & T & T & A & A & - & G & - & A
\end{array}
$$

are two possible alignments of S_1 and S_2. A bit of terminology will come useful later:

> ℹ️ **Gap**
>
> A *gap* is a consecutive series of one or more indel signs.

Therefore, alignment A_1 has two gaps of length 1, while A_2 has one gap of length 2 and two of length 1.

> ℹ️ **Global and local alignments**
>
> An alignment of S_1 and S_2 is *global* if the two sequences appear in their entirety. A *local* alignment is one where only a portion of S_1 is aligned with a portion of S_2.

For example, A_1 and A_2 are global alignments of S_1 and S_2, while

$$
A_3 = \begin{array}{ccccc}
1 & 2 & 3 & 4 & 5 \\
A & T & T & - & A \\
A & T & T & A & A
\end{array}
$$

is a local alignment of the same sequences. We will consider only alignments in which the indel sign is permitted, called *gapped* alignments, although *ungapped* alignments are sometimes considered.

> ℹ **Matches, mismatches, and indels**
>
> Positions in a pairwise alignment in which the same letter appears in the two rows are called *matches*. Those in which two different letters are on top of each other are called *mismatches*. Those in which a letter and an indel sign appear are called *indels*.

1.2.2 Parsimony and Score

For the time being, let us assume that the similarities we are seeking are due to descent from a common ancestor. Then each alignment can be interpreted as representing the evolution of the two sequences from a common ancestor sequence, that we will call S_A, through mutations. Consider alignment A_1: Positions 1, 3, 4, 6, and 8 are matches. We interpret them as saying that the ancestral sequence has not changed in these positions, and S_1 and S_2 have both retained the ancestral base. Position 7 is a mismatch, which we interpret as marking a *substitution* that happened in one of the two lineages leading from S_A to S_1 and S_2. That is, either S_A had a C that was replaced by a G in the lineage leading to S_2; or S_A had a G that was replaced by a C in the S_1 lineage (we cannot tell which of the two actually happened). Finally, indels (positions 2 and 5) represent *insertions* or *deletions* (explaining the name of the symbol "-"). For example, in position 2, either an A was present in S_A and was removed by a deletion in the S_2 lineage, or S_A read "$GTT\ldots$" and an A was inserted in the S_1 lineage; again we do not know which of the two happened.

Clearly, there are many more series of events that could have led from an ancestral sequence to S_1 and S_2. For example, S_A could very well have had a C in the first position that changed into a G in both the S_1 and S_2 lineages. However, our interpretation is the *most parsimonious* one, that is, explains the alignment by postulating the smallest possible number of events. Since usually we have no way of determining the true S_A, and all the mutation events need to be postulated without validation, we prefer an interpretation involving the minimum possible number of unverifiable assumptions. In our parsimonious interpretation, alignment A_1 explains the descent of S_1 and S_2 from S_A by postulating one substitution and two insertion/deletion events.

This concept of parsimony is at the basis of the *scoring* of alignments, a quantitative way to determine which of several possible alignments of the same two sequences is preferable. Consider the alignment A_2 of the same two sequences: The evolutionary process represented by A_2 involves three substitutions and three insertion/deletion events (one of them involving two consecutive letters), and is thus less parsimonious than that represented by A_1. Therefore, we consider A_2 to be inferior to A_1 as an alignment of S_1 and S_2.

This intuition can be made quantitative by defining a *scoring system*, that is, a rule to assign a score to each alignment based on how parsimonious it is. The simplest way of doing this consists in assigning a positive score to each match, and a negative score to each mismatch and each indel. For example, we can assign a score of $+1$ to each match, a score of -1 to each mismatch, and a score of -2 to each indel (assigning a more negative score to indels than mismatches reflects the biological assumption that insertion/deletions are rarer than substitutions). Thus,

$$score(A_1) = 5 \cdot (+1) + 1 \cdot (-1) + 2 \cdot (-2) = 0$$

$$score(A_2) = 2 \cdot (+1) + 3 \cdot (-1) + 4 \cdot (-2) = -9$$

so that our scoring system confirms our intuition that A_1 is a better alignment than A_2.

A few remarks are in order:

- Our discussion was based on the assumption that the sequences to be aligned are the descendants of a common ancestor through mutation processes. While this approach makes it easier to understand the principles behind alignment scoring, sequence alignments are used also when the similarity of the sequences is not due to common ancestry, but, for example, to *convergent evolution*, in which selective pressure forces two unrelated sequences to become similar.

- The scoring system we described is adequate for the alignment of DNA and RNA sequences. More sophisticated scoring systems take into account the fact that some substitutions are more likely than others. Taking this into account becomes indispensable when aligning amino acid sequences (see section 1.2.3).

- We have assigned no meaning to the score itself, but only to the comparison of scores. Thus we established that A_1 is better than A_2, but we have not attempted to use the score to determine "how good" is each alignment in absolute terms. Statistical methods can be used to this purpose, essentially based on determining how likely it would be to obtain a given score from random sequences. We will briefly return to this point when we discuss BLAST (basic local alignment search tool) in section 1.2.6.

- The scoring of insertions/deletions presented here, in which each indel sign is assigned the same (negative) score, is a bit simplistic. It is reasonable to assume that, in the evolutionary process, the enlargement of an existing gap is a more common event than the creation of a new gap. Thus, some scoring systems assign a score to the opening of a new gap (e.g., -2) and a different, less negative score to the enlargement of an existing gap (e.g., -1). In practice, this means that each gap is given a score of -2, and each indel in a gap besides the first one is given a score of -1. Thus, the two alignments of $TAAAAT$ and $TAAT$:

$$
\begin{array}{ccccccc}
& 1 & 2 & 3 & 4 & 5 & 6 \\
A = & T & A & A & A & A & T \\
& T & A & - & A & - & T
\end{array}
$$

$$
\begin{array}{ccccccc}
& 1 & 2 & 3 & 4 & 5 & 6 \\
A' = & T & A & A & A & A & T \\
& T & A & - & - & A & T
\end{array}
$$

which would have the same score of 0 in our previous system, have now different scores, A scoring 0 but A' scoring 1 (four matches, one gap opening, and one gap enlargement). This is called an *affine* gap scoring, while the scoring in which each indel is penalized the same irrespective of its position is called *linear*.

1.2.3 Scoring Protein Alignments

So far we have assigned the same score to all matches and all mismatches, irrespective of the actual letters involved. Implicitly, this assumes that all substitutions are equally likely. However, this is known not to be true in nature: For example, *transitions* (between two purines or two pyrimidines, i.e., $A \leftrightarrow G$ and $C \leftrightarrow T$) happen more often than *transversions* in which a purine changes into a pyrimidine or vice versa. A better scoring system would penalize more the substitutions less likely to occur. Indeed such systems have been developed and are sometimes used in the alignment of nucleotide sequences.

When aligning sequences of amino acids, this problem cannot be ignored, since the occurrence rates of the various possible amino acid substitutions are so different that a scoring system penalizing them equally would be wholly inadequate. Differences in substitution rates stem from two main factors: (1) amino acids represented by similar codons are more likely to replace one another and (2) substitutions involving amino acids with similar physico-chemical properties are more likely to be tolerated by natural selection and thus to be observed in nature. We then base our scoring on a suitable *substitution matrix*:

> **i Substitution matrix**
>
> A *substitution matrix* specifies the score to be assigned to each possible match and mismatch. It is therefore a symmetric square matrix with as many rows and columns as there are possible letters in the sequences to be aligned (4 for nucleotide and 20 for amino acid sequences).

Substitution matrices for protein sequences have been developed based on how often each match or mismatch is encountered in actual proteins. Therefore, substitutions occurring more often are given higher (less negative) scores than rarer ones, and some very common substitutions can even have a positive score. The BLOSUM62 substitution matrix is among the most commonly used in the alignment of amino acid sequences. It is the 20×20 matrix shown below:

	C	S	T	A	G	P	D	E	Q	N	H	R	K	M	I	L	V	W	Y	F
C	9	−1	−1	0	−3	−3	−3	−4	−3	−3	−3	−3	−3	−1	−1	−1	−1	−2	−2	−2
S	−1	4	1	1	0	−1	0	0	0	1	−1	−1	0	−1	−2	−2	−2	−3	−2	−2
T	−1	1	5	0	−2	−1	−1	−1	−1	0	−2	−1	−1	−1	−1	−1	0	−2	−2	−2
A	0	1	0	4	0	−1	−2	−1	−1	−2	−2	−1	−1	−1	−1	−1	0	−3	−2	−2
G	−3	0	−2	0	6	−2	−1	−2	−2	0	−2	−2	−2	−3	−4	−4	−3	−2	−3	−3
P	−3	−1	−1	−1	−2	7	−1	−1	−1	−2	−2	−2	−1	−2	−3	−3	−2	−4	−3	−4
D	−3	0	−1	−2	−1	−1	6	2	0	1	−1	−2	−1	−3	−3	−4	−3	−4	−3	−3
E	−4	0	−1	−1	−2	−1	2	5	2	0	0	0	1	−2	−3	−3	−2	−3	−2	−3
Q	−3	0	−1	−1	−2	−1	0	2	5	0	0	1	1	0	−3	−2	−2	−2	−1	−3
N	−3	1	0	−2	0	−2	1	0	0	6	1	0	0	−2	−3	−3	−3	−4	−2	−3
H	−3	−1	−2	−2	−2	−2	−1	0	0	1	8	0	−1	−2	−3	−3	−3	−2	2	−1
R	−3	−1	−1	−1	−2	−2	−2	0	1	0	0	5	2	−1	−3	−2	−3	−3	−2	−3
K	−3	0	−1	−1	−2	−1	−1	1	1	0	−1	2	5	−1	−3	−2	−2	−3	−2	−3
M	−1	−1	−1	−1	−3	−2	−3	−2	0	−2	−2	−1	−1	5	1	2	1	−1	−1	0
I	−1	−2	−1	−1	−4	−3	−3	−3	−3	−3	−3	−3	−3	1	4	2	3	−3	−1	0
L	−1	−2	−1	−1	−4	−3	−4	−3	−2	−3	−3	−2	−2	2	2	4	1	−2	−1	0
V	−1	−2	0	0	−3	−2	−3	−2	−2	−3	−3	−3	−2	1	3	1	4	−3	−1	−1
W	−2	−3	−2	−3	−2	−4	−4	−3	−2	−4	−2	−3	−3	−1	−3	−2	−3	11	2	1
Y	−2	−2	−2	−2	−3	−3	−3	−2	−1	−2	2	−2	−2	−1	−1	−1	−1	2	7	3
F	−2	−2	−2	−2	−3	−4	−3	−3	−3	−3	−1	−3	−3	0	0	0	−1	1	3	6

As expected, substitutions between similar amino acids are less penalized than those between dissimilar ones. For example, all substitutions involving aromatic amino acids (Dayhoff class "f", namely W, Y, and F) have a positive score. The biological explanation is that these substitutions are unlikely to radically modify the three-dimensional structure, and thus the function, of the protein, and are thus well tolerated by natural selection. Note, however, that the matrix itself is not built by considering the physico-chemical properties of the amino acids, but is based exclusively on the observed substitution frequencies.

Gaps in amino acid sequence alignments are usually scored with an affine scoring system, that is, by assigning different penalties to the opening of a new gap and to the extension of an existing one. For example, the default in the commonly used BLAST alignment software (see section 1.2.6) is to assign a score of −11 to the opening and −1 to the extension of a gap. Thus, for example, the following alignment:

1	2	3	4	5	6	7	8	9	10	11	
W	G	K	V	G	A	H	A	G	E	Y	
W	G	K	V	-	-	N	V	D	E	V	
+11	+6	+5	+4	−11	−1	+1	0	−1	+5	−1	=18

has a score of 18 when evaluated with the BLOSUM62 substitution matrix and the gap scoring system mentioned above.

1.2.4   Alignment Algorithms

Our first task when comparing two sequences is therefore to find their best possible (i.e., highest-scoring) alignment. In principle one could simply generate all the possible alignments, compute their scores, and choose the highest. In practice, this is possible only for very short sequences. Indeed, the number of possible alignments of two sequences grows very fast with their length: For example, for two sequences of equal length N, the number of possible alignments is given by the formula:

$$\binom{2N}{N} = \frac{(2N)!}{(N!)^2}$$

which becomes huge even for moderate values of N: For example, there are $\sim 10^{119}$ possible alignments of two sequences of $N = 200$ nucleotides[2]. Fortunately, *alignment algorithms* have been developed that allow us to obtain the highest scoring alignment of two sequences in a reasonable time. Alignment algorithms can be classified based on whether they deal with global or local alignments and whether they provide an exact or approximate result.

> ℹ **Global and local alignment algorithms**
>
> - *Global* alignment algorithms provide the best global alignment of two sequences S_1 and S_2, that is, the best scoring alignment of the entire S_1 with the entire S_2.
>
> - *Local* alignment algorithms produce the best local alignment of the sequences, that is, find among all possible subsequences of S_1 and all possible subsequences of S_2 the two that can be aligned with the maximum score, and provide such alignment.

Whether to seek a global or a local alignment depends on whether we are interested in assessing similarities along the whole length of the sequences (as in the case of the descendants of an ancestral gene in two extant species), in which case we need a global alignment; or in comparing unrelated sequences that might contain regions of similarity (such as specific functional domains), which can be identified by a local alignment.

---

[2]For comparison, the total number of protons in the observable universe, the so-called Eddington number, is estimated to be $\sim 10^{80}$, that is, 39 orders of magnitude smaller. This implies that listing all possible alignments of two sequences of $N = 200$ nucleotides is simply not possible with any conceivable technology.

Moreover, alignment algorithms can be classified as *exact* or *heuristic*:

---

i **Exact and heuristic alignment algorithms**

An alignment algorithm is *exact* if it can be mathematically proven that it provides the highest-scoring alignment of the input sequences; it is *heuristic* if it provides an alignment that is with high probability similar or identical to the highest-scoring one.

Heuristic alignment algorithms typically use a fraction of the computational resources needed by exact algorithms.

1.2.5 Exact Alignment Algorithms

Exact algorithms are based on *dynamic programming*, a technique in which a complex problem (finding the best alignment of two sequences S_1 and S_2) is broken down into simpler sub-problems that are solved recursively (finding the best alignment of subsequences of S_1 and S_2 of increasing length). The *Needleman-Wunsch* and *Smith-Waterman* algorithms are exact algorithms based on dynamic programming able to find, respectively, the best global and local alignment of two sequences.

As an example of a global alignment, consider the microRNA[3] miR-298, and specifically its mature sequence, which miRBase reports as made of 24 nucleotides. This microRNA has been recently implicated in Alzheimer disease [5] and its sequence in various mammals is available from the same database. We will compare the human mature miR-298 sequence with those of the same microRNA[4] in the macaque (*Macaca mulatta*) and the mouse through global alignments using the Needleman-Wunsch algorithm. We will use the scoring system introduced before for nucleotide sequences, which assigns +1 to a match, −1 to a mismatch, and a linear penalty of −2 to each indel.

The algorithm tells us that the best possible alignment with the macaque miRNA sequence has 22 matches, 2 mismatches, and no indels, for a total score of:

$$(+1) \cdot 22 + (-1) \cdot 2 + (-2) \cdot 0 = 20$$

The alignment is shown below:

```
macaque 1  AGCAGAAGCCGGGUGGUUCUCCCA      24
           |||||||| ||| |||||||||||
human   1  AGCAGAAGCAGGGAGGUUCUCCCA      24
```

---

[3]microRNAs are short RNA molecules able to bind the 3′ UTRs of mRNAs and regulate their decay and translation. The *mature* microRNA sequence has the typical length of 20–25 nucleotides, and derives from a *precursor* sequence of length ∼80. miRBase is a database of microRNA sequences and annotations, and reports for each microRNA both the mature and the precursor sequence.

[4]More precisely, of the *orthologs* in macaque and mouse of the human miR-298. See section 1.5.2, for a precise definition of orthology.

Here we are using a format in which the numbers before and after each sequence are the positions in the input sequence of the aligned portion (which is the whole sequence for global alignments, but not for local ones), and matches are indicated by vertical bars in the middle row. When we do the same using the mouse miR-298 sequence (of length 23, as reported by miRBase) instead of the macaque one we obtain the following global alignment:

```
mouse    1 GGCAGAGG-AGGGCUGUUCUUCCC      23
           ||||| | |||| ||||| ||
human    1 AGCAGAAGCAGGGAGGUUCUCCCA      24
```

with 17 matches, 6 mismatches, and 1 indel, for a total score of 9. As expected, the score of the human/mouse alignment is lower than that of the human/macaque one because more time has elapsed since the divergence between human and mouse than between human and macaque, so there has been more time available to accumulate mutations. We will see below that this is the central idea allowing us to reconstruct how a set of sequences have descended from a single common ancestor[5].

As an example of local alignment, we will consider two small regions of the human genome (one of length 32 on chromosome 18 and one of length 42 on chromosome 12) that bind the same transcription factor (MYC), as experimentally shown by ChIP-seq (see Chapter 5) in human embryonic stem cells. When we attempt a global alignment (with the Needleman-Wunsch algorithm), the score is very low (−6), and the two sequences seem completely unrelated for most of their length:

```
site_chr18  1 GG---GAAGG--GGT--TCTCA-TCCCTG--ATGCACGTGGC    32
              ||   | ||| ||   || || |       |  | |||||||||
site_chr12  1 GGTTTGGAGGAAGGCAATCACAGTTGGAGGGAAGCACGTGGA    42
```

Indeed, there is no reason to believe they descended from the same ancestral sequence. However, since they both are able to bind MYC, they should have in common the DNA sequence that is recognized by MYC. Since transcription factors typically recognize sequences quite shorter that the ones we are considering, a local alignment could reveal small subsequences with high similarity. In fact, the Smith-Waterman algorithm finds that the best possible local alignment of the two sequences selects a subsequence of length 11 from each, which can be aligned with a score of 9:

---

[5] A note of caution: In this example indeed the number of differences follows the pattern expected from the known evolutionary relationships between the three species. This is by no means true for all genes, for various reasons that have to do with differences in selective pressure and/or mutation rates between the lineages, or to more complex phenomena such as incomplete lineage sorting (see, e.g., the Wikipedia page "Incomplete lineage sorting"). In general, inferences about the evolutionary relationships between species must be drawn from the comparison of many genomic regions.

```
site_chr18   21  GATGCACGTGG      31
                 || |||||||||
site_chr12   31  GAAGCACGTGG      41
```

Both aligned subsequences contain the short sequence known to be bound by MYC ("*CACGTG*").

1.2.6   A Heuristic Alignment Algorithm: BLAST

One of the most common use cases of alignment algorithms is the following: Given a sequence, find whether similar sequences have been deposited in a large sequence database. This is useful, for example, to infer the function of a sequence based on its similarity with other sequences of known function. In this case, one has to perform as many pairwise alignments as there are sequences in the database of interest. For example, the GenBank database contains more than 1.5 billion nucleotide sequences. Therefore, even if, on a modern workstation, the exact alignment algorithms described before take only a few hundredths of a second to perform a pairwise alignment of two sequences ranging in the hundreds of base pairs, searching for a matching sequence in GenBank using these algorithms is not feasible.

However, *heuristic* algorithms have been developed that trade accuracy for speed: Unlike exact algorithms, they do not guarantee (in a mathematically provable way) that the alignment they return is the highest scoring one, but perform the alignments at much greater speed allowing searches on huge databases. By far, the most commonly used heuristic algorithm is BLAST (basic local alignment search tool [1]), which performs local gapped pairwise alignments. While BLAST can be used to perform a single pairwise alignment, its advantages are especially evident when a single sequence of interest (the *query*) is compared to a database of *target* or *subject* sequences.

This is achieved through a *preprocessing* of the query, whose main steps are:

- make a list of all words of length k that are found in the query. k is a tunable parameter: In the NCBI implementation of BLAST for nucleotide sequences, the default is $k = 11$. Thus, the human miR-298 sequence considered above:

$$AGCAGAAGCAGGGAGGUUCUCCCA$$

produces the words

```
AGCAGAAGCAG
 GCAGAAGCAGG
  CAGAAGCAGGG
   AGAAGCAGGGA
    GAAGCAGGGAG
    ...
```

- for each word, list its *neighbors*, that is, all the words that would obtain a score at least equal to a pre-established minimum when aligned with it. For example, using our previous scoring system, and a minimum score of 8, the words *ACCAGAAGCAG* and *AGCUGAAGCAG* would be included in the neighbors of *AGCAGAAGCAG* as they would give a score of 9 (10 matches and 1 mismatch).

Simplifying a bit, exact matches to the neighbors are then located in the sequences of the target database (using a highly efficient matching algorithm) and each match is used as a possible *seed* for the alignment. The alignment itself is obtained by expanding in both directions from the seeds. The key point is that a lot of work is done on the query sequence, and hence only once, before looking at the target database, so the gain in execution time is especially significant when the query needs to be aligned to a large database.

However, aligning a query to a large database creates a new problem, because, if the database is sufficiently large, high-scoring alignments can appear simply by chance. To quantify this effect, BLAST outputs for each alignment an *E-value* (expect score).

> **i BLAST *E*-value**
>
> The *E*-value of an alignment is the number of hits with score equal to or better than that of the alignment, that we would expect to find when aligning our query to a database of sequences *unrelated* to it, and of the same size as the database used.

Thus, a small *E*-value suggests that the alignment found is probably biologically meaningful, since it is unlikely to have appeared by chance, also when taking into account the size of the database. This type or reasoning is related to the problem of *multiple testing* that we will treat in some detail in Chapter 2.

If we take the human miR-298 sequence and use BLAST to search for matches in the RefSeq database[6], restricting the search to the *M. mulatta* RNA sequences, the macaque miR-298 is the top hit (i.e., the best-scoring alignment), with an *E*-value of 0.47, as shown below:

```
>Macaca mulatta microRNA mir-298 (MIR298), microRNA
Sequence ID: NR_032462.1 Length: 88
Range 1: 11 to 34

Score:32.2 bits(16), Expect:0.47,
Identities:22/24(92%),  Gaps:0/24(0%), Strand: Plus/Plus

Query  1    AGCAGAAGCAGGGAGGTTCTCCCA   24
```

---

[6]RefSeq is an annotated collection of RNA and DNA sequences with reduced redundancy with respect to the larger Nucleotide database

```
                    ||||||||| ||| |||||||||
Sbjct   11   AGCAGAAGCCGGGTGGTTCTCCCA   34
```

A few comments on this result:

- Note that BLAST always translates RNA sequences into the corresponding DNA sequence (*U*s become *T*s).
- The sequence for the macaque miR-298 in the RefSeq database is 88-nucleotide long, much longer than our query: This is because our query is the mature human miRNA sequence, while RefSeq contains the whole miRNA precursor. However, BLAST is a *local* alignment algorithm, and thus extracted from the macaque precursor sequence, the portion that best aligned with the human mature sequence.
- The alignment is exactly the same we obtained using the Needleman-Wunsch algorithm when we performed a global alignment of the two mature sequences. Such agreement between heuristic and exact methods, in general, is not guaranteed, and will not happen for longer or more divergent sequences.
- The score is not the same we obtained, since this alignment was obtained with BLAST's default scoring system for short query sequences, which differs from the system we used so far. In particular, matches are given a score of +1 and mismatches to −3, so that the score is 16 (shown in parentheses). The reported score in bits (32.2) derives from a more complex scoring system used in particular to compute the *E*-value.

The *E*-value of 0.47 could cast some doubt on the significance of the result: It implies that a database of unrelated sequences, as large as the *M. mulatta* RNA RefSeq database, would be expected to produce, on average, 0.47 alignments with a score at least as good as the one we obtained. This is unavoidable for short sequences (for which, intuitively, it is easier to find good matches, just by chance, in a large database of unrelated sequences) and suggests that this alignment, by itself, would not be enough to conclude that this macaque sequence is indeed derived from the same ancestral sequence as the human one. An obvious solution is to consider a longer sequence, for example, the whole human miRNA precursor, as our query. If we do that we get as the best alignment:

```
>Macaca mulatta microRNA mir-298 (MIR298), microRNA
Sequence ID: NR_032462.1 Length: 88
Range 1: 1 to 88

Score:141 bits(156), Expect:3e-33,
Identities:84/88(95%),  Gaps:0/88(0%), Strand: Plus/Plus

Query  1    TCAGGTCTTCAGCAGAAGCAGGGAGGTTCTCCCAGTGGTTTTCCTTGACTGTGAGGAACT   60
            |||||||||||||||||| ||| ||||||||||||||||||||||||||||||||||||||
Sbjct  1    TCAGGTCTTCAGCAGAAGCCGGGTGGTTCTCCCAGTGGTTTTCCTTGACTGTGAGGAACT   60
```

```
Query  61  AGCCTGCTGCTTTGCTCAGGAGTGAGCT  88
           |||||||||| |||||||||||| ||||||
Sbjct  61  AGCCTGCTGTTTTGCTCAGGAATGAGCT  88
```

whose tiny E-value guarantees that the alignment was not the result of chance, supporting the notion that these sequences indeed derive form a common ancestral sequence.

It would be natural at this point to proceed to multiple alignments, but we need to make a digression and discuss phylogenetic trees, as one of the most used algorithms for multiple alignment, which we will discuss in some detail, uses such reconstruction as a preliminary step.

1.3 PHYLOGENETIC TREES

1.3.1 Speciation and Phylogenetic Trees

Changes in the DNA sequence due to replication errors, when happening in the germline, lead to differences in the genomes of the members of a population, which we refer to as *genetic variation*. When the gene flow between two populations of the same species is interrupted (e.g., because of a geographical barrier forming between them), genetic differences accumulate until they are so large that the two populations are not cross-fertile anymore[7], and become two different species. A series of such *speciation* processes leading from a common ancestor to a number of extant species is described by a specific type of graph called a *phylogenetic tree*. For example, the graph in Figure 1.1 represents the likely sequence of events leading from a common ancestor to seven extant species of great apes. In computational biology, phylogenetic trees are used in many contexts, not necessarily related to speciation, as we will see later in this chapter and in Chapter 3. However, the descent of species from a common ancestor is conceptually the simplest context for introducing the subject.

In the figure, each circle represents a species, either extant (named species on the right) or postulated to be an ancestor of some of the extant species. The axis represents evolutionary time (in million years). The tree tells us that, about 20 million years ago (Mya) the common ancestor of the great apes underwent a speciation event. One lineage eventually led to orangutans (genus *Pongo*, currently including the species *P. pygmaeus* and *P. abelii*), while the other led to the African apes: humans (today represented by *H. sapiens*), chimps (the genus *Pan* with the two extant species *P. troglodytes*, the chimpanzee, and *P. paniscus*, the bonobo), and gorillas (genus *Gorilla*, with the extant species *G. gorilla* and *G. beringei*). About 10 Mya, the common ancestor of the African apes split into a lineage that eventually led to gorillas, and one leading to humans and chimps, and so on.

In most cases, we do not have direct access to data about the ancestral species. Instead, we need to infer the phylogenetic tree based on data on the extant species only. The purpose of this section is to show how such inference is performed. First, we need to introduce some mathematical terminology, whose purpose is to precisely

---

[7]More precisely, cannot produce fertile offspring when crossed.

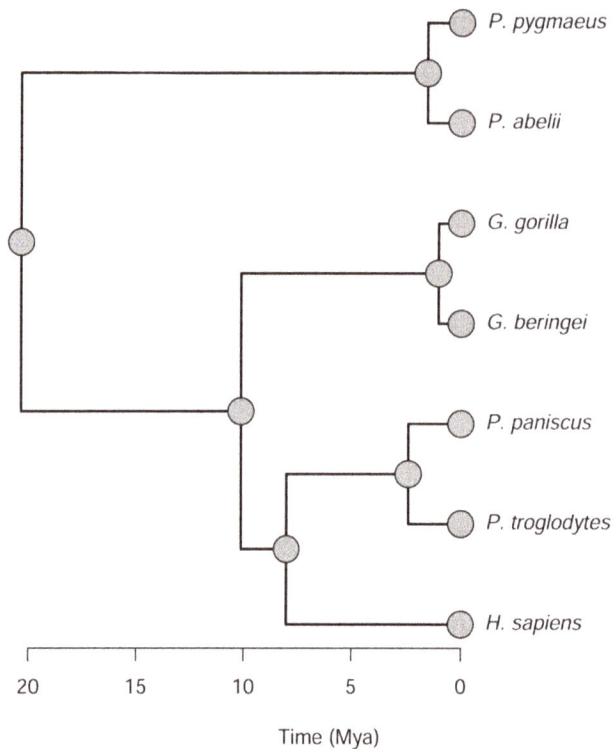

Figure 1.1 A graph showing the sequence of speciation events leading from a common ancestor living ~20 million years ago to seven extant species of great apes. Each circle represents a species, either extant (named species on the right) or postulated to be an ancestor of some of the extant species. The axis represents evolutionary time (in million years).

define the types of graphs that we will use to describe known or inferred evolutionary histories.

1.3.2 Graph Theory Terminology

> **i Graph**
>
> A *graph* is defined by a set of *nodes* and a set of *edges*. Each edge connects two nodes. If a starting node and an ending node are specified for each edge, the graph is called *directed*, otherwise it is *undirected*.

Graphically, we will represent nodes as circles and edges as lines, or arrows in the case of directed graphs. For example, Figure 1.2A shows an undirected graph with seven nodes and seven edges.

We need to define some properties of graphs, whose meaning is actually quite intuitive. We will not give formal mathematical definitions (that are easily found in Wikipedia, e.g., in the page "Path (graph theory)").

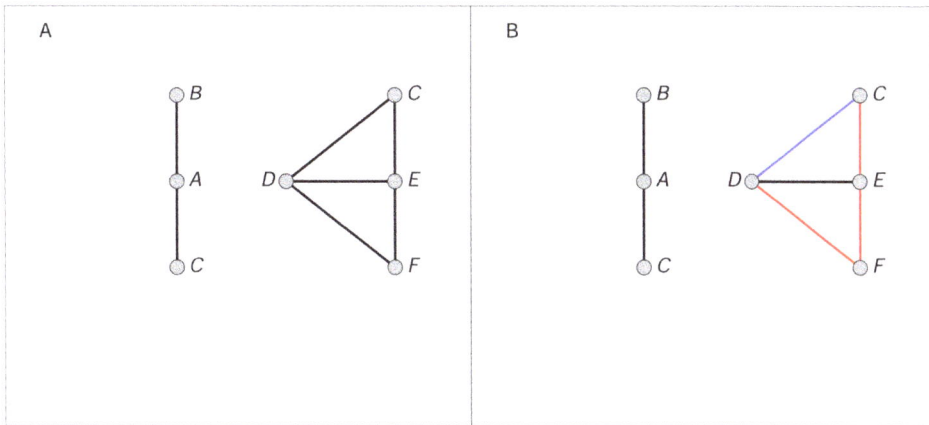

Figure 1.2 (A) An undirected graph with seven nodes and seven edges. (B) A path made of one edge (blue) and one made of three edges (red) between nodes C and D.

> **i** **Path**
>
> A *path* from node X to node Y is a set of consecutive edges leading from X to Y.

Thus, in our previous graph, we have, among others, a path made of one edge (blue) and one made of three edges (red) between nodes C and D (Figure 1.2B).

> **i** **Connected graph**
>
> A graph is said to be *connected* if for any two nodes there is at least one path between them.

Thus the graph in Figure 1.2 is not connected, but the one we obtain by adding an edge from A to D is connected (Figure 1.3A).

> **i** **Tree**
>
> A *tree* is a graph in which each pair of nodes is connected by *exactly one* path.

It follows in particular that all trees are connected. The graph in Figure 1.3A is not a tree because, for example, D and E are connected by three paths. The graph shown in Figure 1.3B is a tree.

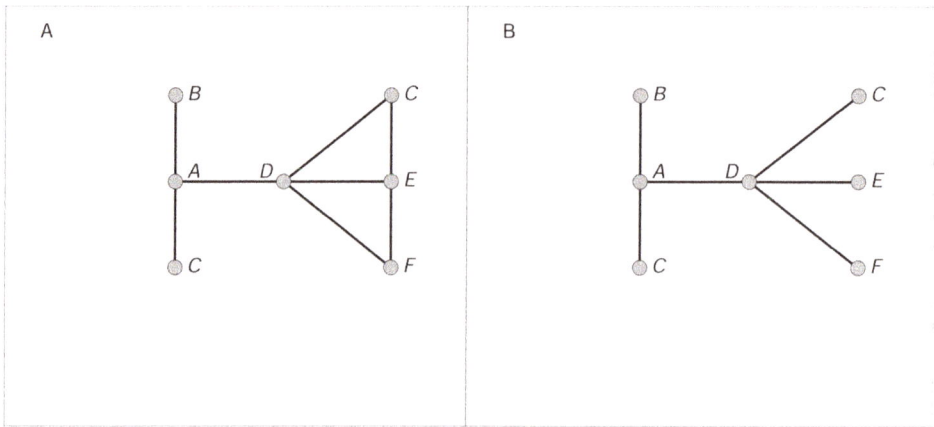

Figure 1.3 (A) A connected graph. (B) A tree.

> **i Unrooted binary tree**
>
> An *unrooted binary tree* is a tree in which every node is attached to either one or three edges. Nodes attached to one edge are called *leaves*, the others are the *internal nodes*.

Finally,

> **i Rooted binary tree**
>
> A *rooted binary tree* is a tree in which all the nodes behave as in an unrooted one (i.e., they are attached to one or three edges) except for one (and only one) special node, the *root*, which is attached to two edges.

Figure 1.4 shows an unrooted binary tree (A) and a rooted binary tree (B).

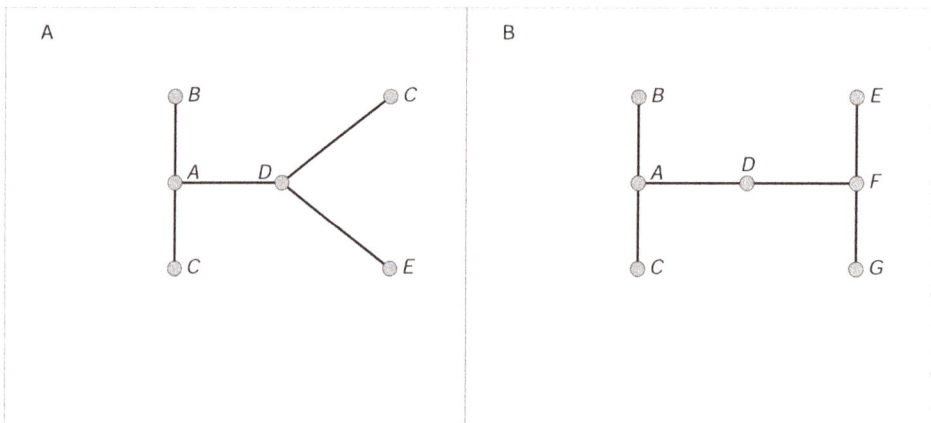

Figure 1.4 (A) An unrooted binary tree. (B) A rooted binary tree, with node D as the root.

Going back to Figure 1.1, we recognize that the graph we used to represent the evolution of great apes is indeed a rooted binary tree. We can now interpret the various ingredients going into the definition of a rooted binary tree in biological terms:

- the *leaves* represent extant species;
- the *root* represents the most recent common ancestor (MRCA) of all the extant species;
- the *internal nodes* represent species descended from the MRCA and which eventually gave rise to the extant species through further speciation events;
- the *length* of each edge is proportional to the time elapsed between two speciation events (or between the last speciation event involving an extant species and the present time, for edges leading to leaves). Note that in the graphical representation used in Figure 1.1, that we will use for phylogenetic trees, the length of an edge is measured on the x axis only; and
- the use of *binary* trees to represent the evolutionary process implies that we assume that speciation always proceeds by generating two (and not more) species from one[8].

1.3.3 The Reconstruction of Phylogenetic Trees

We are interested in reconstructing the phylogenetic tree representing the evolutionary history of a group of related species, that is, the series of speciation events that allowed them to derive from a single common ancestor. In most cases we have few or no *fossil data*, that is, direct information about the ancestral species represented by the internal nodes and the root of our tree. Several methods have been devised to *infer* the tree using only data about the extant species by exploiting their similarities: The general idea is that extant species that are more similar to each other are more likely to be derived from a recent speciation event, as there has been less time available for differences to accumulate.

Distance-based methods explicitly define a quantitative notion of distance between each pair of extant species and use this distance to infer the tree. Other methods are based on the concepts of *parsimony*, that is, attempt to explain the derivation of the extant species from a common ancestor by postulating the smallest possible number of evolutionary changes (an idea that we have already used to justify the scoring of alignments). Finally, the most sophisticated methods are based on statistics through the concept of likelihood. We will describe in detail a distance-based method, UPGMA, because of its simplicity and because it will return later in this chapter and also in Chapter 3 in a quite different context. Note however that modern phylogenetic research tends to use statistical methods that we will mention only briefly.

---

[8]While in principle one could conceive events in which more than two species are generated simultaneously, these could be represented by binary trees in which two nodes are separated by a very short edge: In the limit of vanishing length, this becomes equivalent to a speciation event producing three species simultaneously.

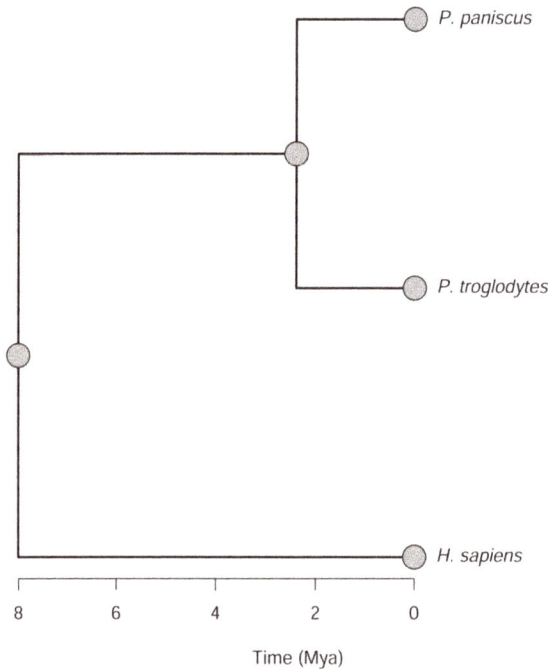

Figure 1.5 The portion of the ape phylogenetic tree depicting the descent of humans and chimps from a common ancestor illustrates in a simple case the concept of evolutionary distance between two extant species (twice the time elapsed from the MRCA of the two species). The evolutionary distance is ultrametric, meaning that given three species, two of the distances are equal and greater than the third one. Here the distance between either *Pan* species and *H. sapiens* is ~16 Myr, and is greater than the distance between the two *Pan* species (~4 Myr).

1.3.4 Evolutionary Distance and Ultrametricity

> **i Evolutionary distance**
>
> The evolutionary distance between two extant species is defined as twice the time elapsed since their most recent common ancestor.

We usually do not know the evolutionary distances between species, since this would require abundant fossil data, but for some well-studied groups of related species a mix of fossil and genetic data allow us to estimate them quite precisely. This is the case, for example, of the great apes shown in Figure 1.1. Let us examine a small part of that phylogenetic tree shown in Figure 1.5.

This tells us that the MRCA of the two extant species of chimp (*Pan*) lived until about 2 Mya, so that the evolutionary distance between the two *Pan* species is ~4 million years (Myr). The reason why we use *twice* the time elapsed from the MRCA is that this is indeed the time of divergence: The differences between the two species had 4 Myr to accumulate (2 Myr in the branch leading to *P. troglodytes* and 2 Myr in the branch leading to *P. paniscus*).

The evolutionary distance can be defined between any two extant species since we believe all species ultimately derive from a single common ancestor. It is a *distance* in the mathematical sense, that is, it shares with the geometrical distance between points in space the following three properties: Given three species A, B, and C, we always have:

- $d(A, B) = 0$ if and only if $A = B$
- $d(A, B) = d(B, A)$
- $d(A, C) \leq d(A, B) + d(B, C)$

Moreover, the evolutionary distance has an additional property that is not shared by the ordinary distance between points, namely, *ultrametricity*:

> ℹ **Ultrametric distance**
>
> A measure of distance is *ultrametric* if given three points A, B, and C, and their three distances, two of them are equal to each other and greater than the third one.

Thus, for example, the ordinary geometric distance is not ultrametric, since otherwise all triangles would be isosceles. On the other hand, it is not difficult to see that the evolutionary distance is indeed ultrametric. For example, for the three species of apes shown in Figure 1.5, the evolutionary distances between *H. sapiens* and the two *Pan* species are the same ($\sim$16 Myr) and greater than the one between the two *Pan* species ($\sim$4 Myr). By referring to the larger tree shown in Figure 1.1, it is easy to convince oneself that this is true for any choice of the three species.

1.3.5 UPGMA

The method for phylogenetic tree reconstruction that we will describe in some detail, UPGMA[9], solves the following mathematical problem:

- Given N objects and a matrix of ultrametric distances between them (*input distances*), find the rooted binary tree which has the objects as the leaves and such that the distances measured on the tree are equal to the input ones.

By "distance measured on the tree" between two nodes, we mean the sum of the lengths of the edges forming the (unique) path joining the two nodes. that particular, UPGMA reconstructs the phylogenetic tree of N extant species from their evolutionary distances as these are ultrametric. The following matrix contains a recent estimate of the evolutionary distances in Myr between the seven apes shown in Figure 1.1 (from [20]):

---

[9]The acronym stands for Unweighted Pair Group Method with Arithmetic mean, which is hardly more informative than the acronym itself

	Pongo pyg-maeus	*Pongo abelii*	*Gorilla gorilla*	*Gorilla beringei*	*Pan troglodytes*	*Pan panis-cus*	*Homo sapiens*
Pongo pygmaeus	0	3.1	40.64	40.64	40.64	40.64	40.64
Pongo abelii	3.1	0	40.64	40.64	40.64	40.64	40.64
Gorilla gorilla	40.64	40.64	0	2.06	20.24	20.24	20.24
Gorilla beringei	40.64	40.64	2.06	0	20.24	20.24	20.24
Pan troglodytes	40.64	40.64	20.24	20.24	0	4.78	16.02
Pan paniscus	40.64	40.64	20.24	20.24	4.78	0	16.02
Homo sapiens	40.64	40.64	20.24	20.24	16.02	16.02	0

UPGMA proceeds recursively as follows:

1. identify the pair of species with the shortest distance;
2. define an ancestral node as the MRCA of these two species and place it on the graph, connected to the two species by edges of length equal to 1/2 of their distance;
3. replace the two species with their MRCA and compute the distance between this MRCA and all other species (see below for how this is done), thus obtaining a distance matrix with $N - 1$ species; and
4. repeat until only one species is left. This species represents the MRCA of the N extant species.

It is much easier to follow in practice: For our $N = 7$ apes, the pair with the smallest distance is *Gorilla gorilla* and *Gorilla beringei*. Thus we will start by defining their MRCA (which we could name *ur-Gorilla*[10]) as an ancestral node placed at time of $2.06/2 = 1.03$ Mya (Figure 1.6A).

Now we need to remove *G. gorilla* and *G. beringei*, replace them with their MRCA, and compute a new distance matrix. The only issue is how to compute the distance between *ur-Gorilla* and the other species. The rule is:

- *The distance between an ancestral node and a leaf is the average of the distances between the leaf and the leaves descending from the ancestral node*

which can be seen as a special case of the general rule

---

[10]"ur-" is a German prefix meaning "primeval."

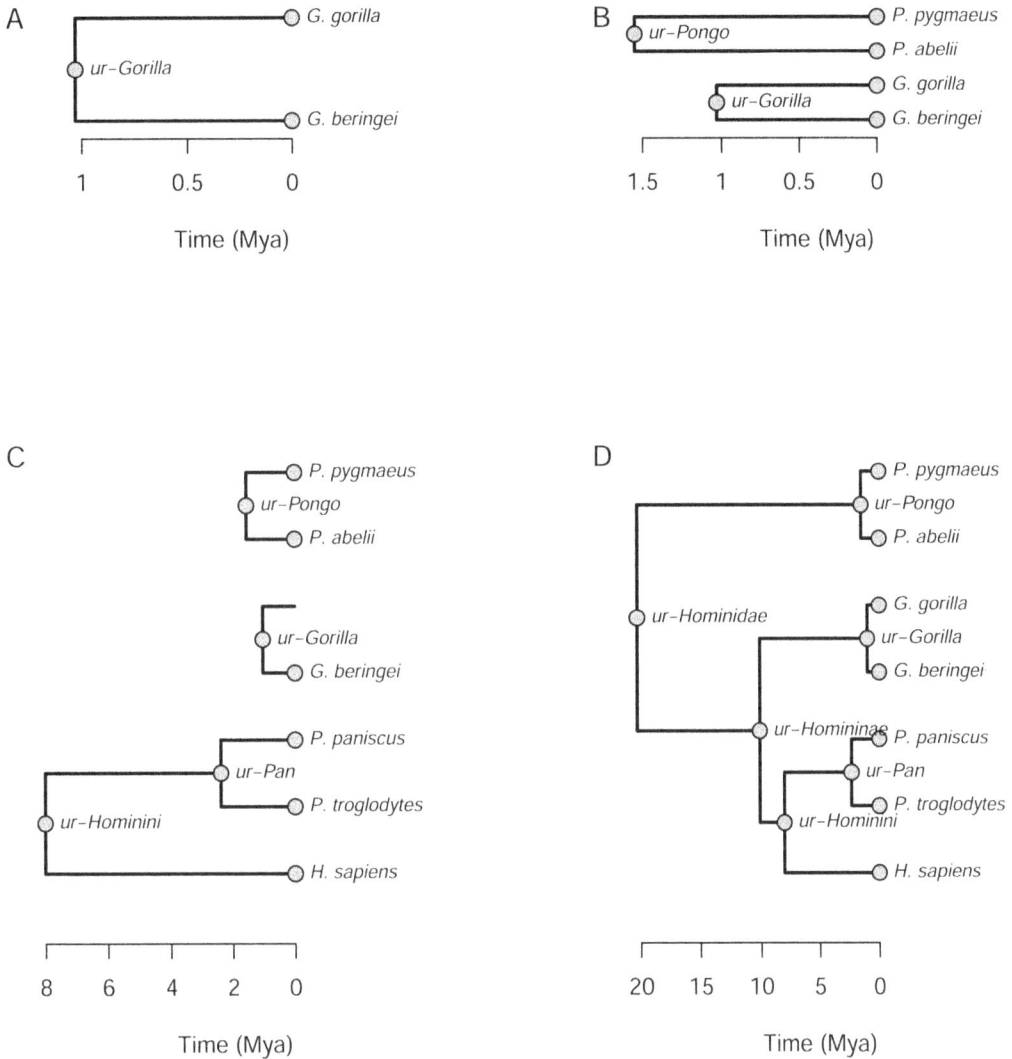

Figure 1.6 Reconstruction of the phylogenetic tree of great apes using UPGMA. (A) First, we create the most recent common ancestor (MRCA) of the pair of species with the shortest evolutionary distance, the two *Gorilla* species; we indicate their MRCA by *ur-Gorilla*. (B) In the next step, we add the MRCA of the *Pongo* genus. (C) After adding the MRCA of the *Pan* genus, the shortest distance in the matrix is between *H. sapiens* and *ur-Pan*. Since human and chimps form the tribe *Hominini*, we call their MRCA *ur-Hominini*. (D) With two more steps we complete the phylogenetic tree.

- *The distance between two ancestral nodes A and B is the average of all pairwise distances between the leaves descending from A and those descending from B*

So, the distance between our newly introduced ancestral node *ur-Gorilla* and, say, *Pan troglodytes* is simply the average of the distances between *Pan troglodytes* and the two *Gorilla* species (which however are equal, because of ultrametricity), and for the next step we have the distance matrix:

	Pongo pygmaeus	*Pongo abelii*	*Pan troglodytes*	*Pan paniscus*	*Homo sapiens*	*ur-Gorilla*
Pongo pygmaeus	0	3.1	40.64	40.64	40.64	40.64
Pongo abelii	3.1	0	40.64	40.64	40.64	40.64
Pan troglodytes	40.64	40.64	0	4.78	16.02	20.24
Pan paniscus	40.64	40.64	4.78	0	16.02	20.24
Homo sapiens	40.64	40.64	16.02	16.02	0	20.24
ur-Gorilla	40.64	40.64	20.24	20.24	20.24	0

By inspection of this matrix, we see that in the next step we need to join the two *Pongo* species with an *ur-Pongo* new ancestral node. So we obtain the next iteration of the tree (Figure 1.6B). Proceeding in the same way, we then remove the two *Pan* species and replace them with their *ur-Pan* ancestor. The shortest distance in resulting matrix is now the one between *Homo sapiens* and *ur-Pan*, so we join them to obtain the partial tree shown in Figure 1.6C and the distance matrix:

	ur-Gorilla	*ur-Pongo*	*ur-Hominini*
ur-Gorilla	0	40.64	20.24
ur-Pongo	40.64	0	40.64
ur-Hominini	20.24	40.64	0

Since the genera *Pan* and *Homo* form together the higher taxonomic unit ("tribe") of *Hominini*, we have called their common ancestor *ur-Hominini*. With two more steps we get to the final tree (Figure 1.6D), which coincides with the one shown in Figure 1.1, to which we have added some more ancestor names: *Homininae* is the subfamily containing the genera *Gorilla*, *Homo*, and *Pan*, that is, the African apes, while *Hominidae* is the family containing all great apes.

The fact that we were able to find the appropriate taxonomic group to name each ancestor is not a lucky coincidence: Modern *cladistic* taxonomy aspires to organize species into *clades* such that each clade contains all and only the descendants of an ancestral species. In this way, there is a one-to-one correspondence between ancestral species (i.e., internal nodes of the phylogenetic tree) and taxonomic groups.

1.3.6 Surrogate Distances

As described, UPGMA reconstructs the phylogenetic tree starting from evolutionary distances. However, evolutionary distances are rarely known, and we need to resort to *surrogate distances*, that we can consider, at least approximately, proportional to the true evolutionary distances. If our surrogate distances are not ultrametric, the distances computed on the tree reconstructed with UPGMA will still be ultrametric, and hence not equal to the input surrogate distances.

For example, if one assumes that mutations accumulate at a constant rate along the whole phylogenesis, the number of sequence differences between two species would be exactly proportional to the time elapsed since their divergence, hence to the true evolutionary distance, and could be used to reconstruct the tree with UPGMA. This assumption is called the *molecular clock* hypothesis.

However, the molecular clock hypothesis rarely holds in practice, for several reasons ranging from differences in mutation rates to the effects of selection. Thus, in many cases, we do not have surrogate distances that we can trust to be approximately proportional to the true evolutionary distances. A sure sign that this is the case is when the surrogate distances are not even approximately ultrametric. In these cases, we need to abandon UPGMA and use other methods, some of which are briefly described below.

1.3.7 Other Methods

Neighbor joining (NJ) is, like UPGMA, a distance-based method for the reconstruction of phylogenetic trees, which, however, does not rely on the ultrametricity of the distances. Trees produced by NJ differ from those produced with UPGMA in two main respects: (1) NJ trees are unrooted (although post-hoc *rooting* of the trees is possible) (2) The length of the branches of NJ trees are proportional to the distance used in input and not to the time elapsed since the split (the two coincide only for ultrametric distances).

Other methods are not based on a notion of distance, but on models of how biological sequences evolve. Given a set of extant sequences, these methods find the most plausible tree that generated the extant sequences from an ancestral one according to the model. The simplest, conceptually, is *maximum parsimony*: The model of sequence evolution is simply that sequences change as little as possible during evolution. Thus, the tree that best describes the evolutionary history of a set of sequences is the one that explains how the extant sequences derived from an ancestral sequence by postulating the smallest possible number of changes along the whole tree (we already encountered this concept of parsimony when we introduced the scoring of sequence alignments). Note that the computational problem of finding the optimal tree is very hard, and cannot be solved exactly except for a small number of extant sequences, so that heuristic methods are usually needed.

Finally, *maximum likelihood* methods are based on probabilistic models describing the evolution of sequences, by specifying the probability of each possible substitution and that of creating or extending a gap. Thus each possible tree describing the derivation of the extant sequences from a common ancestor is associated with a probability based on the evolutionary model, and we choose the tree for which this probability is maximal (again, in practice, this needs to be done with heuristic algorithms).

1.4   MULTIPLE ALIGNMENTS

1.4.1   Scoring Multiple Alignments

Equipped with phylogenetic trees and methods to reconstruct them, we can now return to sequence alignments and tackle multiple alignments. A *multiple alignment* is simply an alignment of more than two sequences, and can be defined similarly to pairwise alignments:

> **i** **Multiple sequence alignment**
>
> A multiple alignment of $N > 2$ sequences is a way of displaying the sequences on top of each other, with the possibility of inserting the *indel* symbol "-" in any position of each sequence, as long as no column contains only indel signs.

Thus, this is a multiple alignment of 4 short DNA sequences:

$$
A_M =
\begin{array}{ccccccccc}
1 & 2 & 3 & 4 & 5 & 6 & 7 & 8 \\
G & A & T & T & - & A & C & A \\
G & - & T & T & A & A & G & A \\
G & A & T & T & - & - & G & A \\
G & A & T & C & A & C & G & A
\end{array}
$$

The score of a multiple alignment can be defined based on a scoring system for pairwise alignments with the *sum of pairs* method.

> **i** **Sum of pairs score**
>
> Given a scoring system for pairwise alignments, the *sum of pairs* score of a multiple alignment of N sequences is the sum of the scores of all the $\frac{N(N-1)}{2}$ pairwise alignments generated by the multiple alignment.

$N(N-1)/2$ is indeed the number of ways in which two different sequences can be extracted from a set of N, disregarding the order in which they are extracted. A pairwise alignment *generated* by the multiple one is simply the one obtained by considering only the two rows corresponding to the sequences of interest. Note that some of the pairwise alignments generated by the multiple one will show indel signs on top of each other, since the definition of multiple alignment only requires that no column contains only indels. For example, this happens for the multiple alignment shown above when we extract the pairwise alignment of the first and the third sequence. Therefore, to our pairwise scoring system, we will need to add a rule to score such pairs of indels.

Let us use a simple scoring method (match $= +1$; mismatch $= -1$; indel $= -2$) for pairwise alignments. We will give a score of 0 to the positions containing two indels[11]. Rather than computing and summing the scores of the 6 pairwise alignments generated by the multiple one, it is perhaps more instructive to compute the score of each column and then sum them[12]. In our example above, the first column has all matches, hence all six possible pairs of letters contribute a score of $+1$, and the score of the column is $+6$. Now consider position 4: Of six pairs of letters that we can extract, three are mismatches (choosing any T and the C) and three are matches (choosing two Ts in three possible ways), and thus this column has a score of 0. The reasoning is similar for column 2, but here we have indels instead of mismatches, thus the score is $3 \cdot (+1) + 3 \cdot (-2) = -3$. For column 6, we have one match, two mismatches, and three indels, summing to -7. Reasoning in the same way we obtain the final score of $+1$:

$$
A_M = \begin{array}{ccccccccc}
1 & 2 & 3 & 4 & 5 & 6 & 7 & 8 \\
G & A & T & T & - & A & C & A \\
G & - & T & T & A & A & G & A \\
G & A & T & T & - & - & G & A \\
G & A & T & C & A & C & G & A \\
\hline
+6 & -3 & +6 & 0 & -7 & -7 & 0 & +6 & = 1
\end{array}
$$

where in column 5 we had one match, 4 pairs with a letter over an indel (the two As combined in all possible ways with the two indels), and one zero-scoring pair featuring two indels.

This scoring system is reasonable but not entirely satisfactory, because if we reason in terms of parsimony, as we did when we introduced the scoring of pairwise alignments, we see that evolutionary events are actually overcounted. For example, consider position 4: The most parsimonious interpretation is that only one substitution happened in the history of these sequences, but the scoring system counts three mismatches as if three events happened. Besides this problem, while exact algorithms for maximizing the SP score, similar in principle to the Needleman-Wunsch and Smith-Waterman algorithms discussed above, are available, they are seldom used in practice since they become computationally prohibitive already for a moderate number of sequences.

1.4.2 Progressive Multiple Alignments

Progressive algorithms are heuristic algorithms for multiple sequence alignment that allow fast alignment of a large number of sequences, although they do not guarantee that the final result is the one with the maximum possible SP score.

---

[11] This is reasonable since the two indels could be simply removed if we were aligning only these two sequences.

[12] This is equivalent to summing the scores of the six pairwise alignments as long as a linear, and not affine, gap penalty is used.

> **i** **Progressive multiple alignment algorithm**
>
> A *progressive* algorithm for multiple alignment works by first aligning two of the N sequences, then adding one sequence at a time to the alignment, until all sequences are aligned.

Two decisions must be made:

(1) the order in which the sequences are aligned;
(2) how to add a new sequence to the existing alignment.

We will describe the Feng-Doolittle algorithm used (with some minor modifications) by the popular CLUSTALW package. As an example, we will seek the multiple alignment of four short DNA sequences using our previous scoring system (match = +1; mismatch = −1; indel = −2). The sequences are:

	sequence
s1	GTTACCCTTG
s2	GATACAATAG
s3	GTATACAATTG
s4	GATTACAAAAT

The first step is to find all the $N(N-1)/2 = 6$ best pairwise alignments between all possible pairs of sequences. Using the Needleman-Wunsch algorithm, we obtain:

```
s1 vs. s2
GTTACCCTTG
| ||| | |     score = 2
GATACAATAG


s1 vs. s3
GT-TACCCTTG
|| ||| |||    score = 4
GTATACAATTG


s1 vs. s4
G-TTACCCTTG
| ||||        score = -2
GATTACAAAAT


s2 vs. s3
G-ATACAATAG
| |||||| |    score = 6
GTATACAATTG
```

```
s2 vs. s4
GAT-ACAATAG
||| |||| |      score = 4
GATTACAAAAT


s3 vs. s4
GTATACAATTG
|  |||||        score = 1
GATTACAAAAT
```

The second step is to create a *guide tree* based on these pairwise alignments. The idea is to consider the score of each alignment as a measure of how closely related the two sequences are, from which we derive a phylogenetic tree with the sequences as the leaves, that will determine the order in which they are progressively added to the alignment. The most appropriate way to proceed is to define a distance as a suitable decreasing function of the score, and use this distance as input to e.g. UPGMA or neighbor joining. In practice, a simpler approach is often taken in which

- for each alignment, we compute the *percent identity*, that is, the ratio of the number of matches to the length of the alignment;
- the distance between two sequences is defined as 1 minus the percent identity.

Thus, in our case the distances are:

	s1	s2	s3	s4
s1	0	0.4	0.2727	0.5455
s2	0.4	0	0.1818	0.2727
s3	0.2727	0.1818	0	0.4545
s4	0.5455	0.2727	0.4545	0

Using UPGMA, we obtain the tree shown in Figure 1.7, which tells us that we have to first align $s2$ and $s3$, then add $s1$ to the alignment, and finally add $s4$.

For the first alignment we simply use the pairwise alignment of $s2$ and $s3$:

```
s2  G-ATACAATAG
s3  GTATACAATTG
```


Then we add $s1$ using as guide its alignment with $s3$, because the score of the pairwise alignment of $s1$ and $s3$ is higher than that of $s1$ and $s2$.

```
s2  G-ATACAATAG
s3  GTATACAATTG
s1  GT-TACCCTTG
```

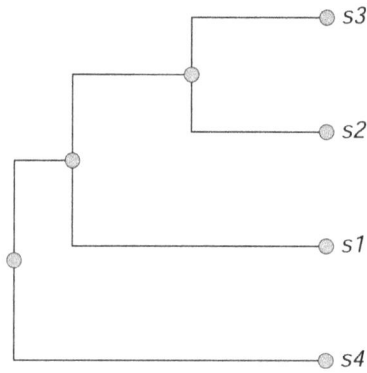

Figure 1.7 The guide tree built from the matrix of pairwise distances between the four sequences to be aligned.

If, instead, we had used the pairwise alignment of $s1$ and $s2$ as a guide, we would have obtained a different alignment (which requires inserting an indel in $s1$):

```
s2   G-ATACAATAG
s3   GTATACAATTG
s1   G-TTACCCTTG
```

which actually is slightly *better* than the previous one in terms of SP score, as you can verify with some patience. It is important to keep in mind that progressive alignment algorithms are heuristic procedures that do not optimize the SP score, or any other score. The complete multiple alignment is obtained by adding $s4$ using its pairwise alignment with $s2$:

```
s2   G-AT-ACAATAG
s3   GTAT-ACAATTG
s1   GT-T-ACCCTTG
s4   G-ATTACAAAAT
```

1.5 COMPARATIVE GENOMICS

1.5.1 Conservation and Function

Multiple alignments provide the foundation of *comparative genomics*, which classifies genomic regions based on their *conservation*, that is, the degree to which they can be found, possibly changed but still recognizable, in other species. Conservation of a region in a set of related species suggests (by the usual parsimony argument) that it was present in their MRCA, and has been propagated with little change through speciation events. This is significant because *functional* regions of the genome, defined as those whose sequence affects the phenotype, and in particular the fitness of the individual, are subjected to selective pressure, which causes individuals with

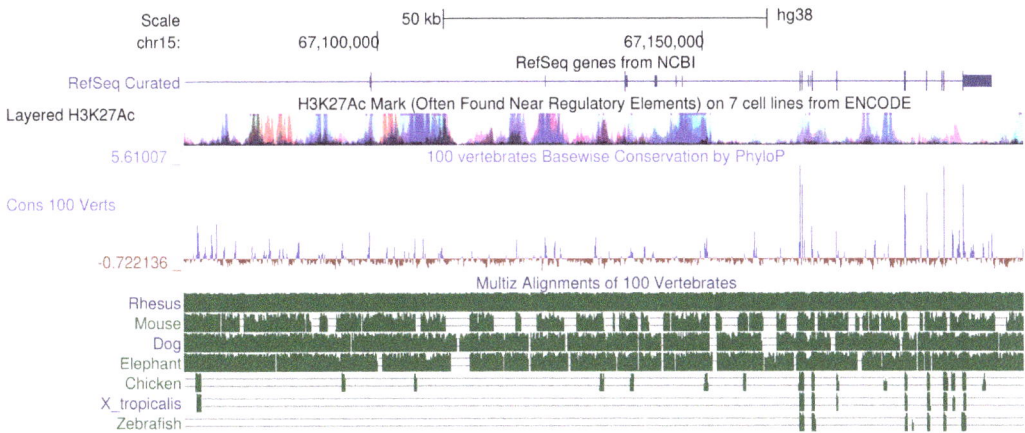

Figure 1.8 A ~130 Kbp region of chromosome 15 displayed in the UCSC genome browser. The three horizontal bands (*tracks*) display three types of genome annotation: "RefSeq Curated" shows the location of exons and introns; "Layered H3K27ac" shows regions characterized by this histone modification, which marks active regulatory regions; "Cons 100 Verts" shows the multiple alignment of this region to the genomes of 100 vertebrates, and the phyloP score representing the degree of conservation at single-base resolution.

mutations in such regions to contribute less, or not at all, to the genomes of later generations. Thus the level of conservation of a genomic region can be used as a proxy for its functional relevance.

Starting with multiple alignments of whole genomes (obtained with algorithms significantly more complex than the one we described above), each individual base of a genome of interest (say the human genome) is assigned a score reflecting its degree of conservation. A popular scoring system is phyloP [31], and the phyloP scores of all bases of the human genome can be displayed in the UCSC genome browser[13]. In the most commonly used version, these scores are based on the multiple alignment of 100 vertebrate genomes, although alignments of 470 mammalian species have recently become available together with their associated phyloP scores. We are not going to discuss in detail how the score is computed; instead, we will consider an example based on a genomic region as displayed by the UCSC genome browser.

Figure 1.8 shows a ~130 kilobase region of human chromosome 15 containing the *SMAD3* gene. Each of the stacked horizontal bands (*tracks*) represent one type of genome annotation: "RefSeq Curated" displays exons (thicker blue lines) and introns (thin blue lines) of known genes, here *SMAD3*; "Layered H3K27Ac" show regions characterized by the acetylation of the lysine residue at position 27 of the histone H3 protein in seven human cell lines (shown in different colors) – this is functionally

---

[13]The UCSC genome browser is an online tool for the visualization of genome sequences and related annotations (e.g., information about the functional role of each genomic region). It is strongly recommended that you become familiar with this free tool, for example, using the excellent tutorials provided by the site. In this text we will only mention what is strictly needed to interpret the figures.

relevant as this modification of H3 is known to be associated with active regulatory regions, as we will discuss in Chapter 5; finally "Cons 100 Verts" shows the phyloP score of each base according to the multiple alignment of 100 vertebrate genomes, and the regions that can be aligned with the genomes of a few representative vertebrate species (green bars). Therefore we can investigate the relationship between *function* (represented by the first two tracks) and *conservation* (third track).

First, we notice that almost all exons correspond to phyloP peaks, implying that exonic regions are more conserved than the rest of the region, as expected since changes in exonic sequence, and especially non-synonymous ones, are much more likely to affect the phenotype, and hence the fitness, than changes within introns. Indeed some exons can be aligned even to the genome of the zebrafish, suggesting that selective pressure has been acting on their sequence for the whole evolutionary history of vertebrates, more than 400 million years. However, there are also conservation peaks within introns, and several of these are located inside active regulatory regions. Since the DNA sequence in regulatory regions affects the binding of transcription factors (see Chapter 5), and thus the expression profile of the gene in tissues and cell types, the conservation of these regions suggests that not only the protein sequence, but also the regulation of $SMAD3$ has been under selective pressure[14].

The alignments shown schematically at the bottom of the figure tell us that the meaning of conservation depends on the evolutionary distances considered: The whole *SMAD3* locus, including all introns, can be aligned with the macaque genome, but this is hardly evidence of selection, since the time elapsed since our last common ancestor could not have been sufficient for the accumulation and selection of enough mutations. On the other hand, comparing the human genome with the zebrafish is likely to identify only the oldest functional regions, and to discard those that have evolved their biological function during the evolution of mammals.

Comparative genomics thus allows us to identify the genomic regions most likely to be functional. Estimates of the fraction of the genome that is under selection vary depending on the methods used to detect it, but most agree that such fraction is around 10% or less. Therefore comparative genomics is a powerful tool allowing us to select genomic regions likely to be biologically important. Of course, on the other hand, not all functional regions are conserved (after all we are neither macaques nor fish), and the search for human-specific, functional genomic regions is important in understanding the genetic origin of human-specific traits.

1.5.2 Homology, Orthology, and Paralogy

We have already used, informally, the term "ortholog" in our discussion of alignments. Here we define orthology more precisely and discuss how orthologous genes can be identified from a phylogenetic tree. We start from a more general definition:

---

[14]We are tacitly assuming that the regulatory regions inside the introns of a gene actually regulate the same gene. This is reasonable but not at all guaranteed, as we will discuss in Chapter 5.

> **i Homologs**
>
> Two genes (in the same species or in different ones) are *homologs* if they have descended from a single ancestral gene.

There are however two distinct processes by which an ancestral gene can produce two descendants. The first is speciation: The ancestral gene resides in a given species, which through the process of speciation gives rise to two species. This is the process represented by the internal nodes of the phylogenetic trees we have discussed above. Each species carries a copy of the gene, and after speciation their sequences will evolve independently, and thus diverge, because of the absence of gene flow between the two new species.

> **i Orthologs**
>
> We call *orthologs* two homologs that separated due to a speciation event. It follows that orthologs always reside in different species.

The other process leading to homology is gene duplication: The duplication of a portion of the genome of a species can give rise to two copies of a gene. These will also tend to diverge in time, especially because selective pressure is now relaxed by the newly acquired redundancy, and thus one of the copies is free to mutate and possibly acquire new functions. If the species undergoes speciation after the duplication event, both new species will inherit the two copies of the ancestral gene.

> **i Paralogs**
>
> We call *paralogs* two homologs that separated due to a gene duplication event. Paralogs can reside in the same or in different species (the latter case requires that a duplication event is followed by one or more speciation events).

In summary, homologs are called orthologs if their split is due to a speciation event, and paralogs if it is due to a duplication event (irrespective of whether the duplication happened in the species where they reside or in any ancestral species). We will now discuss some examples of increasing complexity. Figure 1.9 shows the simplest case: There is only a speciation event in which the gene G_A of the ancestral species A gets split into genes G_1 and G_2 of the derived species 1 and 2, so that G_1 and G_2 are orthologs.

In Figure 1.10, a duplication of gene G_A has taken place in the ancestral species, giving rise to the genes F_A and H_A. After a speciation event in which the ancestral species divides into species 1 and 2, we obtain the four genes F_1, H_1, F_2, and H_2. For each pair of genes, we can follow the branches leading to them backwards in time until they join: If this happens in a speciation (duplication) event, the genes are orthologs (paralogs). Thus, F_1 and F_2 are orthologs, as are H_1 and H_2, while all other pairs of genes are paralogs.

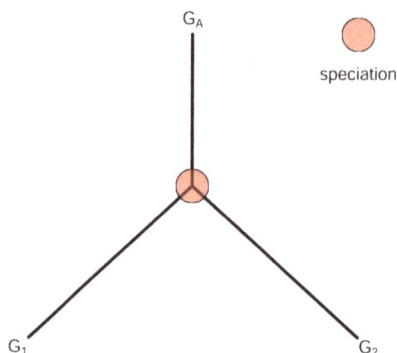

Figure 1.9 Generation of orthologs by speciation: When the ancestral species A divides into the children species 1 and 2, gene G_A gives rise to the orthologous genes G_1 and G_2.

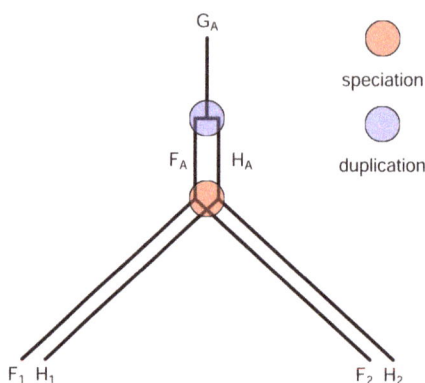

Figure 1.10 Here a duplication of gene G_A has taken place in the ancestral species, giving rise to the genes F_A and H_A. After a speciation event in which the ancestral species divides into species 1 and 2, we obtain the four genes F_1, H_1, F_2, and H_2. For each pair of genes we can follow the branches backward until they merge to determine whether they are orthologs (F_1 with F_2 and H_1 with H_2) or paralogs (all other pairs).

Finally, a more complex situation is shown in Figure 1.11, in which two duplication events have taken place at different times. Following the same rule we conclude, for example, that H_1 has two orthologs (K_2 and J_2) and one paralog (F_2) in species 2.

Thus, in particular, orthology relationships are not necessarily one-to-one, as shown in the last example. Orthologous genes in different species are likely to be involved in similar functions, especially when the relationship is actually one-to-one. The concept of orthology is thus fundamental in comparative genomics, and is at the basis of the use of model organisms for experiments that we cannot conduct in humans, even when our real interest is human biology (think of the experimental investigations of *Trp53*, the mouse ortholog of the human tumor suppressor *TP53*). However, it is important to keep in mind that the *definition* of orthology does not involve function in any way, and is entirely based on the evolutionary history of the genes.

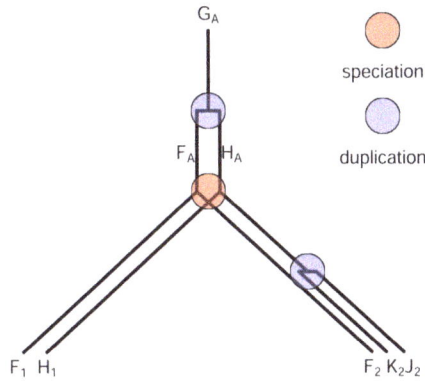

Figure 1.11 A more complex situation with two duplication and one speciation event. H_1 has two orthologs (K_2 and J_2) and one paralog (F_2) in species 2.

1.5.3 Reconstructing Homology Relationships

The definitions given above allow us to determine whether homologous genes are orthologs or paralogs only if we know the complete history of speciation and duplication events that led from an ancestral gene to the genes of interest. Since this is rarely the case, we need a way to do that based only on the sequences of the extant genes. A commonly used method uses *best reciprocal hits*: Genes G_1 and G_2 in species 1 and 2 are deemed to be orthologs if, among all genes in species 2, G_2 is the one that best aligns to G_1, and vice versa. In many practical cases this is sensible, but does not capture the subtleties of the definition of orthology discussed before. In particular, orthology relationships found in this way are always, by construction, one-to-one, while we have seen that this is not necessarily the case.

An elegant method directly inspired by the definitions of orthology and paralogy was introduced in [18] and is based exclusively on *gene trees*, that is, the phylogenetic trees obtained from the sequences of homologous genes in several species. Suppose we do that for the genes shown in Figure 1.11: If the tree reconstruction is correct, we will obtain a tree that has the same topology as the one shown in Figure 1.11, except that we will not be able to distinguish between speciation and duplication events. The gene tree will look like Figure 1.12A.

If we were able to classify the internal nodes of this tree into speciations and duplications, we could use the rules defined above to determine for each pair of extant genes whether they are orthologs or paralogs. This can be achieved by the *species overlap rule*: For each internal node, consider the two branches departing from the node and the leaves (extant genes) descending from each branch. If a node represents a speciation event, these two groups of leaves will contain genes belonging to two non-overlapping sets of species; if the species associated to the two groups overlap, then the node describes a duplication event.

Figure 1.12B shows the result of applying this rule: Node D is a duplication, because species 2 is shared between leaves J_2 and K_2. For node B, the two branches contain H_1 on one side, and J_2 and K_2 on the other, thus the species do not overlap,

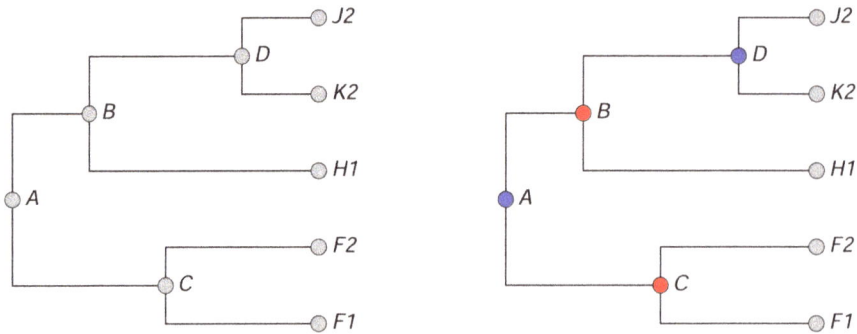

Figure 1.12 (A) Phylogenetic tree built from the genes whose history was shown in Figure 1.11. (B) The same tree where the internal nodes have been classified as duplication (blue) or speciation (red) events using the species overlap rule.

and the node is a speciation event, as is node C, while A represents a duplication event. For each pair of genes, we can now trace back the node where they separated, and classify them as orthologs (paralogs) if such node is a speciation (duplication) event.

1.5.4 An Example: Opsins in Mammals

Opsins are photosensitive proteins responsible for color vision in animals. Most mammals have two opsin genes, *OPN1MW* and *OPN1SW*, respectively, sensitive to medium- (green) and short-wave (blue) light, and thus have dichromatic vision. Old-world monkeys and apes (including humans) have an additional opsin gene, *OPN1LW*, sensitive to long-wave (red) light, allowing us trichromatic vision. Sequence comparison suggests that all these genes are homologs, and we can use the methods discussed above to understand their evolutionary relationships.

Consider the three opsin genes in human and macaque, and the two mouse genes. To understand the evolutionary history of these genes, we can perform a multiple alignment of the corresponding proteins, compute the pairwise distances[15], then build a tree with UPGMA. The tree is shown in Figure 1.13, where we have applied the species overlap rule to classify each node as speciation (red) or duplication (blue).

Thus, a gene duplication gave rise to medium and short-wavelength opsins from an ancestral opsin, then the medium-wave one underwent further duplication in the old-world monkey lineage, one copy becoming the long-wavelength opsin[16]. We can use this tree to assign paralogs and orthologs: For example, all medium- and

---

[15] For example, simply as 1 − percent identity; as mentioned above, there are more sophisticated ways of doing this. However, for this example the results are very robust with respect to the choice of distance measure and phylogenetic tree reconstruction methods.

[16] Regarding the ancestral duplication, this tree only tells us that it happened before the split between rodents and primates. Richer trees involving many vertebrate species suggest that such duplication occurred early in the vertebrate lineage, and also reveal a more complex history with many gene losses.

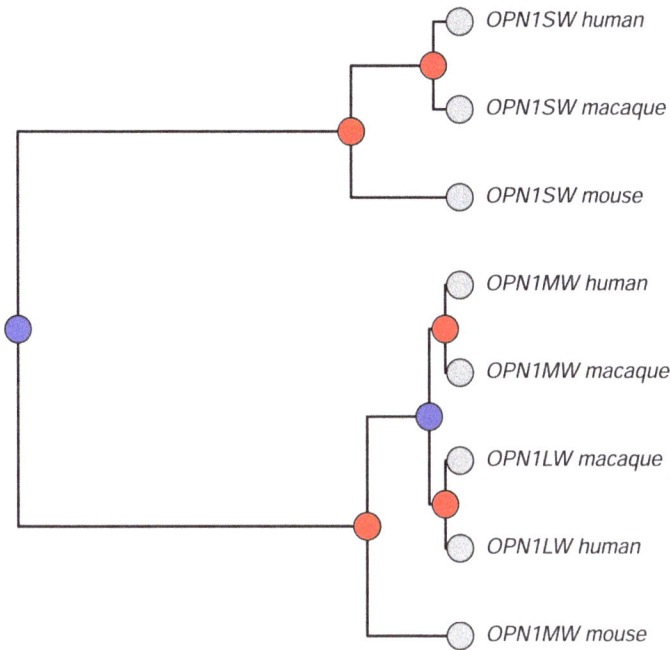

Figure 1.13 Gene tree of the opsins in mouse, macaque, and human, based on their protein sequence. Using the species overlap rule, we can determine which internal nodes are duplications (blue) and which are speciations (red), and thus classify each pair of genes as paralogs or orthologs. For example, all medium- and long-wavelength opsins are paralogs of the short-wave ones; both human *OPN1MW* and *OPN1LW* are orthologs of mouse *OPN1MW*.

long-wavelength opsins are paralogs of the short-wave ones; both human *OPN1MW* and *OPN1LW* are orthologs of mouse *OPN1MW*, etc.

1.6 ANALYSIS OF NEXT-GENERATION SEQUENCING DATA

1.6.1 Reads and the FASTQ Format

We conclude this chapter by discussing the important role of sequence alignment in the analysis of next-generation sequencing (NGS) data[17]. NGS is a set of techniques for sequencing DNA, but as we will see in the following chapters it has revolutionized the entire field of molecular biology, and not only the analysis of the DNA sequence *per se*: Ingenious methods have been developed to use NGS to investigate, for example, gene expression, the binding of transcription factors to regulatory DNA, epigenetic modifications of the genome, and its three-dimensional conformation.

---

[17]"Next generation sequencing" is obviously a rather unfortunate designation (what are we supposed to call the sequencing methods that will come after NGS?), but is quite established and we will use it. "Massive parallel sequencing" is a better name; "Second generation sequencing" is not very expressive but at least leaves room for more generations.

All these methods start with the sequencing of some DNA and the alignment of the resulting sequences to a reference genome[18]. The most popular NGS technologies produce large quantities of *short reads* (typically 50 to 300 bases long) which represent the sequence of fragments of the original DNA molecules. These reads are usually represented in a *FASTQ* file, a text file in which each individual read is represented by four lines of text:

- Line 1 starts with the "@" character and contains an identifier of the individual read and optionally a description.
- Line 2 contains the actual sequence.
- Line 3 starts with the "+" character and sometimes contains again the read identifier, but is often left blank (except for the "+").
- Line 4 contains a quality assessment of each individual base in the sequence, under the form of a string of ASCII characters with each character representing a quality level equal to its numerical ASCII code.

This is an example of a FASTQ file containing 2 reads:

```
@NB501182:29:HGWMCBGXY:1:11101:21691:1066 1:N:0:ATTACTCG+AGGCTATA
GATGCAGAAGAGCAGAAACAGCAGCANNNNCNCCTGGGAGCCGCATCCCCGCGCACATTGAGTGAGGAACCGACCC
+
AAAAAEEEEEEEEEEEAEAAAEEEEEA/####<#E<EEA/EEAAEEEAE/<6A6</EEAEEAAAEEAEEEE<E/EE/E
@NB501182:29:HGWMCBGXY:1:11101:22355:1073 1:N:0:ATTACTCG+AGGCTATA
GCAGTAAGTGGCAAATTAAGAAGTAATCAGAGTTCCATAGGGATAGTCAGGATAGGCTTCCTGGAAGAGGCAGGAT
+
AAAAAEEEEEEEEEEEAEEEEEEEEEEEEEEEEAEEEEEEEEEEEEEEEEEEEEEEEEEEEEEEEEEEEEEEEEEEEEE
```

Note that in the first read some bases could not be identified with confidence, and are shown as "N". The corresponding quality is represented by ASCII symbols with low numerical codes (e.g., "#" correspond numerically to 35 while "E", which appears as the quality of most bases, corresponds to 69. The range of ASCII characters most commonly used to represent quality contains 41 characters from "!"=33 to "I"=73).

1.6.2 Read Alignment

The results of an NGS run are typically FASTQ files containing millions of reads, and the first step of the analysis consists in locating the genomic region from which each read originated. However, we do not necessarily expect the read sequence to coincide exactly with that of a genomic region, as reported in the reference genome because (1) sequencing errors are rare but possible, and, especially, (2) the genome of the individual we are sequencing will not be identical to the reference genome due

---

[18]Or, at least, here we will discuss cases in which the complete genome sequence of the species of interest is available. This is the most common situation, since the genomic sequence of thousands of species is now available. There are however research fields based on sequencing the DNA of species whose complete sequence is unknown, such as, *metagenomics*, which studies the DNA sequences from multiple organisms found in a given environment (such as the human gut, or the ocean).

to genetic variation (the reference human genome is actually a patchwork of the genomes of a few individuals).

Thus we are not seeking, for each read, a perfectly matching genomic region, but rather the best local pairwise alignment of the read with the entire genome: The genomic region most likely to have produced the DNA fragment represented by the read is the one that gives the best alignment. Thus, in principle, we could use BLAST, or even the Smith-Waterman algorithm, to associate each read to a genomic region. However, compared with the examples of pairwise alignment discussed above, the alignment of NGS short reads to the genome has some peculiarities, which led to the development of alignment algorithms specifically tailored to this problem. Without describing these algorithms, we briefly discuss such peculiarities.

First, the alignment problem is highly asymmetric: We are trying to align millions of different short sequences (the reads) to the same huge sequence (the genome). This reminds us of the preprocessing of the query performed by BLAST. In a similar way, NGS-specific alignment algorithms perform an *indexing* of the genome, completely independent of the reads. Second, since the reads are short, each of them could align well to multiple genomic regions. This problem is made especially serious by the fact that the genome of higher eukaryotes abounds in sequences identically or almost identically repeated hundreds of times in many genomic loci. Obviously, a read coming from one of these regions will align well to all the other copies of the region. In this case, we need the algorithm to report all the alignments scoring better than a given threshold, rather than only the absolute best one. Finally, some (but not all) NGS-specific alignment algorithms also take into account the sequencing quality reported in the FASTQ file, by weighing more the bases with higher quality.

1.6.3 Visualizing Aligned Reads

The results of the read alignment are reported in text files following the SAM (Sequence Alignment/Map) format (or, more often, its binary compressed version, the BAM [Binary Alignment Map] format). These files are composed of a *header section* followed by an *alignment section*. The header section contains some general information about how the alignments are shown, and the list of the sequences to which the reads were aligned (typically the chromosomes of the reference genome). The *alignment section* uses one line for each alignment, each column containing information about the read and the result of its alignment to the reference genome, such as:

- the read identifier
- the read sequence
- the read quality (these three columns simply reproduce information contained in the FASTQ file)
- the reference sequence (chromosome) to which the read was aligned
- the coordinate in the reference sequence where the alignment starts

The following is an example of a the alignment section of a SAM file (we show only selected columns and we truncate sequence and quality):

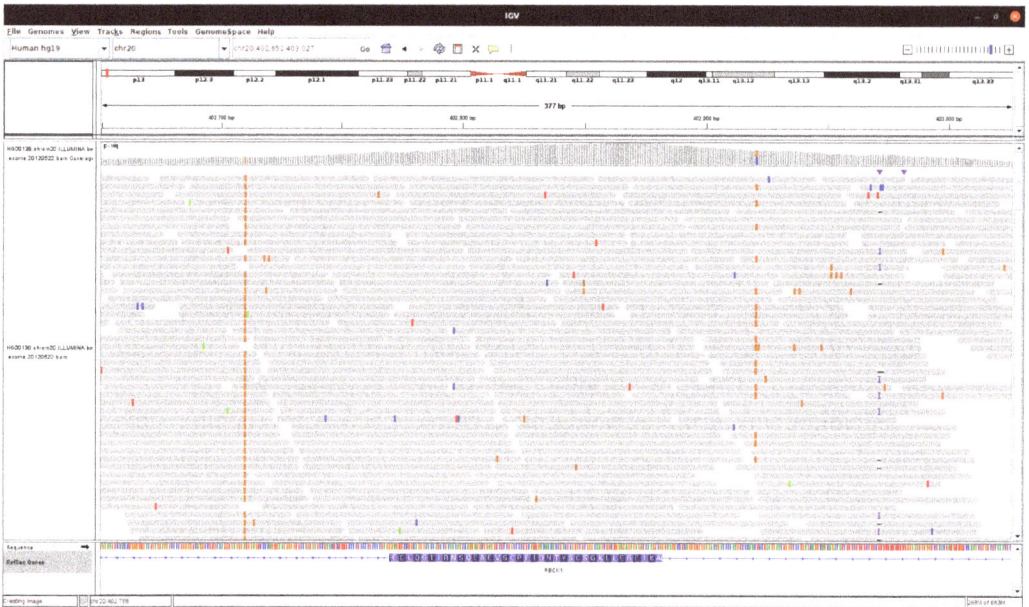

Figure 1.14 Aligned NGS reads in a 377-bp genomic region around an exon of the *RBCK1* gene, shown in the Integrative Genomics Viewer. Differences between the read sequence and the reference genome sequence are shown in color. Many differences appear in isolated reads, and are thus probably sequencing errors. Differences that appear in all reads can be interpreted as homozygous variants (see, e.g., the variant near coordinate 402,700), and those appearing in about half the reads as heterozygous variants (e.g. the one near coordinate 402,900).

```
NS500400:132:HG3VMBGXX:2:12109:13860:1647   chr1    3000811 AAGATCAGCC...   EEAEEAE/E/...
NS500400:72:HCJL5BGXX:4:23403:6260:9196 chr1    3000877 TTGAGGACTT...   AEEE/EEEEE...
NS500400:132:HG3VMBGXX:2:23308:7888:11353   chr1    3001174 ACATTCAGAA...   /EEEEEEE/A...
NS500400:72:HCJL5BGXX:1:12211:5986:9754 chr1    3001250 AACTTTGAGC...   EEEEEEEEEE...
NS500400:72:HCJL5BGXX:3:11401:12312:12449   chr1    3001250 AACTTTGAGC...   EEEEEEEEEE...
NS500400:132:HG3VMBGXX:2:22204:24648:2112   chr1    3001600 TAGTTTCTGT...   AEE//AEEEE...
NS500400:132:HG3VMBGXX:1:12205:7415:12945   chr1    3001726 AGTTTGGCTG...   AAAAAEEEEE...
NS500400:72:HCJL5BGXX:4:22512:11870:13125   chr1    3001968 ATTTGGAATT...   AAAAAEEEEE...
NS500400:72:HCJL5BGXX:4:22507:15572:12342   chr1    3003560 CTGCAGTCTG...   EEEEEEEEEE...
NS500400:72:HCJL5BGXX:1:21211:16231:17134   chr1    3004110 TAAGATAAAA...   AAAAAE6EEE...
```

The key information here is the association between each read and a genomic region: For example, the first read produced a good alignment with the genomic region starting at coordinate 3,000,811 on chromosome 1. The information contained in this file can be visualized with specific programs such as the Integrative Genomics Viewer (IGV): A genomic region is shown (Figure 1.14) as a horizontal line, together with, from top to bottom

- the *coverage* of each base, that is, the total number of reads that have been aligned to a genomic sequence containing that base;

- the reads, represented by rectangles placed exactly where they align; and

- genomic annotation: In Figure 1.14 the location of genes is shown in a way similar to that used by the UCSC genome browser, with exons shown as thick lines and introns as thin ones.

Here, we are looking at a 377 bp region containing an exon of the *RBCK1* gene. This comes from the sequencing of a specific individual (in this case, one of the subjects sequenced for the 1000 Genomes Project), and in general we expect the DNA sequence not to be exactly identical to that of the reference genome.

1.6.4 Variants

The individual bases of the reference sequence are shown by four colors in the "Sequence" track, and differences between the reference sequence and the read sequence are shown using the same colors on the reads. These differences can be classified based on their recurrence and interpreted as follows:

- Many differences appear in isolated reads, and are probably due to sequencing errors;
- Differences appearing in all or almost all reads overlapping a given position likely represent a true difference between the DNA that has been sequenced and the reference genome. For example, one such variant can be seen near coordinate 402,700; and
- The DNA that is sequenced comes with equal probability from the paternal and maternal copies, so the differences that appear in about half the reads can be interpreted as heterozygous variants with respect to the reference genome (one can be seen near coordinate 402,900).

When sequencing DNA from human subjects, we are often interested precisely in the variants carried by each subject, for example, because these could explain inherited phenotypic differences and diseases (see Chapter 6). Thus many statistical methods have been devised to "call" such variants, that is, to distinguish them in a statistically controlled way from sequencing errors. While these methods are often sophisticated, the basic principles are the ones discussed above. In particular, it is clear that the possibility to distinguish true variants from sequencing errors, and homozygous from heterozygous variants, strongly depends on the sequencing *coverage*, that is, the number of reads aligned to the base of interest.

FURTHER READING

Alföldi, J. & Lindblad-Toh, K. Comparative genomics as a tool to understand evolution and disease. *Genome Res.* **23**, 1063–1068 (2013).

Christmas, M., Kaplow, I., et al. Evolutionary constraint and innovation across hundreds of placental mammals. *Science.* **380**, eabn3943 (2023).

Durbin, R., Eddy, S., Krogh, A. & Mitchison, G. Biological Sequence Analysis (Cambridge University Press, 1998).

Koonin, E. Orthologs, paralogs, and evolutionary genomics. *Annu Rev Genet.* **39**, 309–338 (2005).

Kuderna, L., Gao, H., et al. A global catalog of whole-genome diversity from 233 primate species. *Science.* **380**, 906–913 (2023).

Muzzey, D., Evans, E. & Lieber, C. Understanding the basics of NGS: From mechanism to variant calling. *Curr Genet Med Rep.* **3**, 158–165 (2015).

Pollard, K., Hubisz, M., Rosenbloom, K. & Siepel, A. Detection of nonneutral substitution rates on mammalian phylogenies. *Genome Res.* **20**, 110–121 (2010).

Reinert, K., Langmead, B., Weese, D. & Evers, D. Alignment of next-generation sequencing reads. *Annu Rev Genom Hum Genet.* **16** pp. 133–151 (2015).

CHAPTER 2

Transcriptomics, Part I: Class Comparison

2.1 INTRODUCTION

The human body is made of hundreds of cell types, remarkably different from each other in their phenotypes. However, all of them share the same genome. Such variety of outcomes from a single genome is made possible by *gene regulation*, a complex series of mechanisms that allow different cell types to *express* different genes in different amounts. The next four chapters are dedicated to gene regulation, one of the most active fields of research in genomics. We will first discuss methods to analyze the end result of gene regulation, namely, gene expression, and later (chapter 5) we will turn our attention to the molecular mechanisms allowing cells to regulate gene expression.

Each of the many steps that the cell must take to turn a gene into an active protein is subject to regulation: the transcription of the gene DNA, the processing of the primary RNA transcript to build the final mRNA, its export out of the nucleus, the possible degradation of the mRNA molecule, its translation, and possibly the activation of the protein, for example, through phosphorylation.

A protein-coding gene can truly be said to be expressed in a given cell type if the protein is synthesized and active. However, measuring the abundance of proteins, or active proteins, at the scale of the whole proteome, notwithstanding remarkable recent progress, is still much more difficult and expensive than measuring of the abundance of mRNA. Therefore, the vast majority of the studies of gene expression assess the *transcriptome*, that is, the mRNA abundance of all the genes in a biological sample.

Transcriptomics was made possible around the turn of the millennium with the introduction of microarrays, which have been largely replaced today by RNA-sequencing. Since most transcriptomic assays measure mRNA abundance in the cytoplasm, what we are actually observing is the combined result of three regulatory mechanism, namely the regulation of transcription, of RNA processing and export, and of mRNA degradation in the cytoplasm. The first of these steps, transcriptional regulation, is the one with the largest impact on the transcriptome, and in Chapter 5, we will discuss some of the experimental assays that allow us to understand the related molecular mechanisms.

Most journals require the transcriptomic data produced for a publication to be made freely available to the community in public *repositories*, one of the most well known being the Gene Expression Omnibus. The size of this repository is truly impressive, with almost 7 million samples (i.e., individual transcriptomic assays) available as of January 2024[1]. This is useful for two reasons: First, it allows other researchers to re-analyze the data to check whether the conclusions drawn by the authors are justified. Second, and more important, these data contain a huge amount of information that has not been explored by the original authors, who often performed the assay to answer a specific biological question. For example, when studying a specific gene, we might wonder in what types of human cancers it is expressed. This question can be easily answered by accessing a large amount of publicly available cancer transcriptomic data and re-analyzing them to answer our specific question: The information is there, even it has never been extracted.

The transcriptome depends not only on the tissue or cell type but also on many other factors, including, for example, the age of the organism, treatment with drugs, environmental factors, disease, etc. While a transcriptomic assay performed on a single condition can be useful by itself (e.g., to provide a catalog of the genes expressed in a given tissue or cell type), much more biologically useful information can be derived by comparing the transcriptomes of different conditions. For example, we can learn something about cancer by comparing the transcriptomes of cancerous vs. normal tissues, or about development by comparing the transcriptomes at different embryonic stages.

In this chapter, we will discuss transcriptomic assays in which samples belonging to two or more pre-defined *classes* are compared to determine which genes are *differentially expressed*, that is, show different expression levels among the classes. This type of analysis is called *class comparison*. As we will see the problem is complicated by the unavoidable variability in the measured levels of gene expression, even when the experimental conditions are carefully kept identical, which makes it necessary to use replicate experiments analyzed with statistical methods. A gene will be identified as differentially expressed among sample classes if we can be confident that the differences in expression that we observe are not due to the intrinsic variability of the measurement but rather to true differences in gene regulation. In the course of the chapter, we will introduce several statistical concepts that are useful in many aspects of computational genomics, including *hypothesis testing, multiple testing, linear regression*, and *likelihood*.

2.2 DIFFERENTIAL EXPRESSION

The discussion of class comparison for microarray data is technically simpler than for RNA-seq data, while the main concepts are unchanged. Therefore, we will start with microarray data, and we will use as case study a dataset obtained in [3] to

---

[1]To be precise, the Gene Expression Omnibus contains also samples from other types of assays, but the vast majority are indeed transcriptomic assays.

study gene expression changes in the brain of aging zebrafish. Class comparison for RNA-seq data is discussed below in Section 2.4.

The data are deposited in the Gene Expression Omnibus under accession GSE53430 and correspond to the brain transcriptomes of 12 animals, six young ones and six old ones. Each age group is represented by three female and three male animals. We will ignore sex for the time being (see below section 2.3) and divide the 12 samples in two classes based on the age group.

Microarrays use *probes* designed to match the sequence of known genes and to hybridize the cDNA derived from the sample mRNA. The hybridization is then quantified, and after some preliminary analytical steps that do not concern us here, we obtain for each probe a real number, in principle proportional to the quantity of mRNA matching the probe in the biological sample (more precisely, its relative quantity with respect to the total quantity of mRNA in the sample).

When combining the assays for our 12 samples, we obtain a data matrix with 15,617 rows (the number of probes on the specific microarray used here) and 12 columns (the biological samples). This is how the data look like (we show just six probes and three samples):

probe_id	GSM1293346	GSM1293347	GSM1293348
AFFX-Dr-acta1-3_at	8.37	7.838	8.114
AFFX-Dr-acta1-5_at	6.29	6.274	6.352
AFFX-Dr-acta1-5_x_at	7.073	6.784	7.141
AFFX-Dr-acta1-M_at	7.07	5.935	6.578
AFFX-Dr-GAPDH-3_at	13.08	13	13.23
AFFX-Dr-GAPDH-5_at	12.57	11.89	12.62

Note that the data are in $\log_2$ scale, that is, they are proportional to the $\log_2$ of the mRNA abundance.

To make sense of these data, we need a correspondence between probes and genes, which is given as a table with 13,120 rows and 2 columns:

probe_id	Symbol
AFFX-Dr-acta1-3_at	actc1b
AFFX-Dr-acta1-5_at	actc1b
AFFX-Dr-acta1-5_x_at	actc1b
AFFX-Dr-acta1-M_at	actc1b
AFFX-Dr-GAPDH-3_at	gapdhs
AFFX-Dr-GAPDH-5_at	gapdhs

Some genes are associated to more than one probe, while each probe is associated to a single gene: Indeed, we have only 10,582 different gene symbols in the table. Note also that the table containing the gene expression data has more rows than the table with the probe/gene correspondence: Some probes are not associated to any gene, possibly because they were designed on legacy databases of expressed sequences, not all of which correspond to genes in the current annotation of the genome.

We also need a table telling us the class to which each sample belongs:

Sample	Age	Sex
GSM1293346	Young	Male
GSM1293347	Old	Female
GSM1293348	Young	Male
GSM1293349	Young	Male
GSM1293350	Old	Male
GSM1293351	Young	Female
GSM1293352	Old	Male
GSM1293353	Old	Female
GSM1293354	Young	Female
GSM1293355	Old	Female
GSM1293356	Young	Female
GSM1293357	Old	Male

Now we want to find the probes (and hence the genes) that are differentially expressed between old and young animals. The simplest thing to do is to compute the *logarithmic fold change* (logFC) of each probe.

> **i** **Logarithmic fold change**
>
> The *logarithmic fold change* (logFC) of a probe is the difference between the mean logarithmic expression of the probe in the two classes of replicates to be compared.

Comparing old vs. young animals, for the first few probes we obtain:

probe_id	Symbol	logFC
AFFX-Dr-acta1-3_at	actc1b	−0.5604
AFFX-Dr-acta1-5_at	actc1b	−0.2421
AFFX-Dr-acta1-5_x_at	actc1b	−0.2746
AFFX-Dr-acta1-M_at	actc1b	−0.5856
AFFX-Dr-GAPDH-3_at	gapdhs	−0.09805
AFFX-Dr-GAPDH-5_at	gapdhs	−0.05659

A logFC > 0 (<0) means that the gene is, on average, more expressed in old (young) animals. Thus, the results shown above suggest that the genes *actc1b* and *gapdhs* have higher expression in young than in old animals. In this way, we could classify each probe as up- or downregulated in old animals. More practically, we could require a minimum value of the logFC to consider a probe as differentially expressed, for example, by considering as upregulated in old animals all probes with a logFC

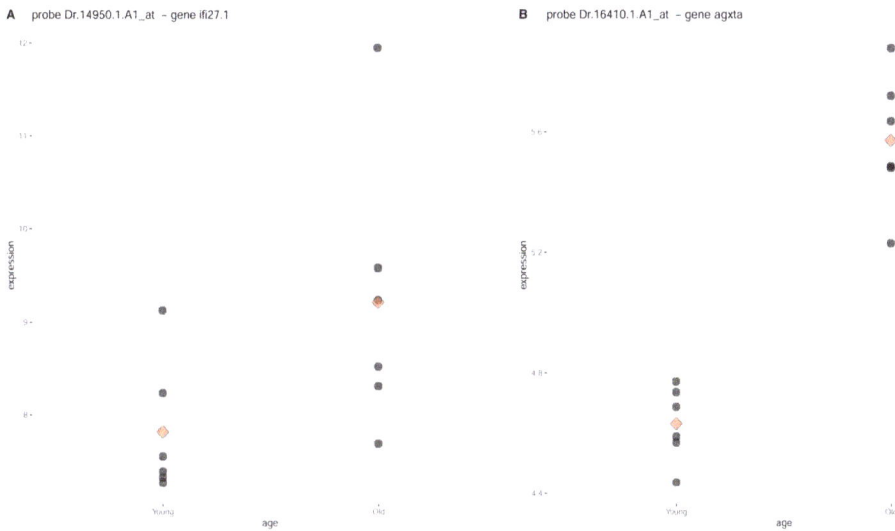

Figure 2.1 (A) Expression level of the gene *ifi27.1*, represented by the microarray probe *Dr.14950.1.A1_at*, in young and old animals. Gray dots are the individual measurements, and red diamonds represent the mean expression of the two classes. The logFC is 1.394, which by itself would suggest that the gene is upregulated in old animals. However, the individual samples, especially for old animals, show very large variability in the measured expression. This variability, rather than true changes in gene regulation, could be responsible for the difference observed. (B) Same for gene *agxta*, represented by the probe *Dr.16410.1.A1_at*. While the logFC is smaller (0.9406), we are intuitively more confident in the fact that this gene is truly differentially expressed, since the variability within each age class is much smaller than that of *ifi27.1*.

≥ 1 (i.e., remembering that we are taking the difference of data expressed in $\log_2$ scale, with at least twice as much mRNA in old animals compared with young ones).

Let us look at an example of such a gene: *ifi27.1*, represented by the probe *Dr.14950.1.A1_at*. The logFC of this probe is 1.39. Its expression in the 12 samples is shown in Figure 2.1A, where the gray dots represent the expression in the individual samples and the red diamonds the mean expression in each age group. The plot shows that the expression of this probe is quite variable among samples, and the upregulation in old animals seems to be mainly driven by a single animal with very high expression. Such variability makes it difficult to conclude from these data that the gene is truly upregulated in old samples.

On the other hand, Figure 2.1B shows the probe *Dr.16410.1.A1_at*, corresponding to the gene *agxta*, with a logFC of 0.941. Although the fold change is smaller, we are intuitively more confident in considering this gene as differentially expressed, since the expression is much more stable among replicates in both age groups.

Variability between replicate experiments is unavoidable, even with the most careful experimental protocols, and is due to both biological variability and uncontrollable technical factors. Replicates are necessary to assess variability, and the use of the mean fold change logFC must be replaced with a procedure that takes variability into account, namely *hypothesis testing*.

2.2.1 Hypothesis Testing

Hypothesis testing is a very general procedure used to extract information from data with inherent variability.

---

i **Hypothesis testing, null hypothesis, and test statistic**

In hypothesis testing, the *null hypothesis*, H_0, usually describes the situation in which nothing of interest happens, so that interesting cases (*discoveries*) are those in which we can *reject H_0*. To do that, we compute a *test statistic*, that is, a number computed from the data whose probability distribution is known when H_0 is true. Values of the statistic that are unlikely under H_0 suggest rejecting it

---

A bit more terminology will be useful later: The logical negation of the null hypothesis is called the *alternative hypothesis* and is often indicated with H_1. Therefore by rejecting H_0, we show that H_1 is true.

Turning to our specific problem, for each microarray probe, the null hypothesis is that the "true" expression level is the same in old and young samples[2], and the difference we see in the plot is due to random fluctuations. To test this null hypothesis, we can use the t-statistic defined by:

$$t = \frac{\bar{X}_O - \bar{X}_Y}{\sqrt{\frac{s_O^2 + s_Y^2}{n}}}$$

where $\bar{X}_O$ is the mean expression in old samples, s_O the corresponding standard deviation, and similarly for the young samples; $n = 6$ is the number of replicates (if the number of replicates is not the same in the two conditions, the definition of t changes slightly). Note that the numerator is just the logFC; however, the denominator depends on the standard deviations of the samples in the two age groups, and becomes large when either class shows high variability among the replicates. Intuitively, a gene with a large t (in absolute value) has a large logFC and/or low variability among replicates, and is thus a gene that we can confidently consider differentially expressed between the two classes.

For example, the values of the t statistic for the probes shown above and associated to the genes *ifi27.1* and *agxta* are, respectively, 2.05 and 8.97, suggesting that we are much more confident in the differential expression of *agxta* than that of *ifi27.1*, in agreement with the intuition we derive from looking at the expression plots.

A bit more formally, if we assume

a. that the null hypothesis is true;

---

[2]The "true" expression value can be defined as the mean expression value that we would obtain from an arbitrarily large number of replicates.

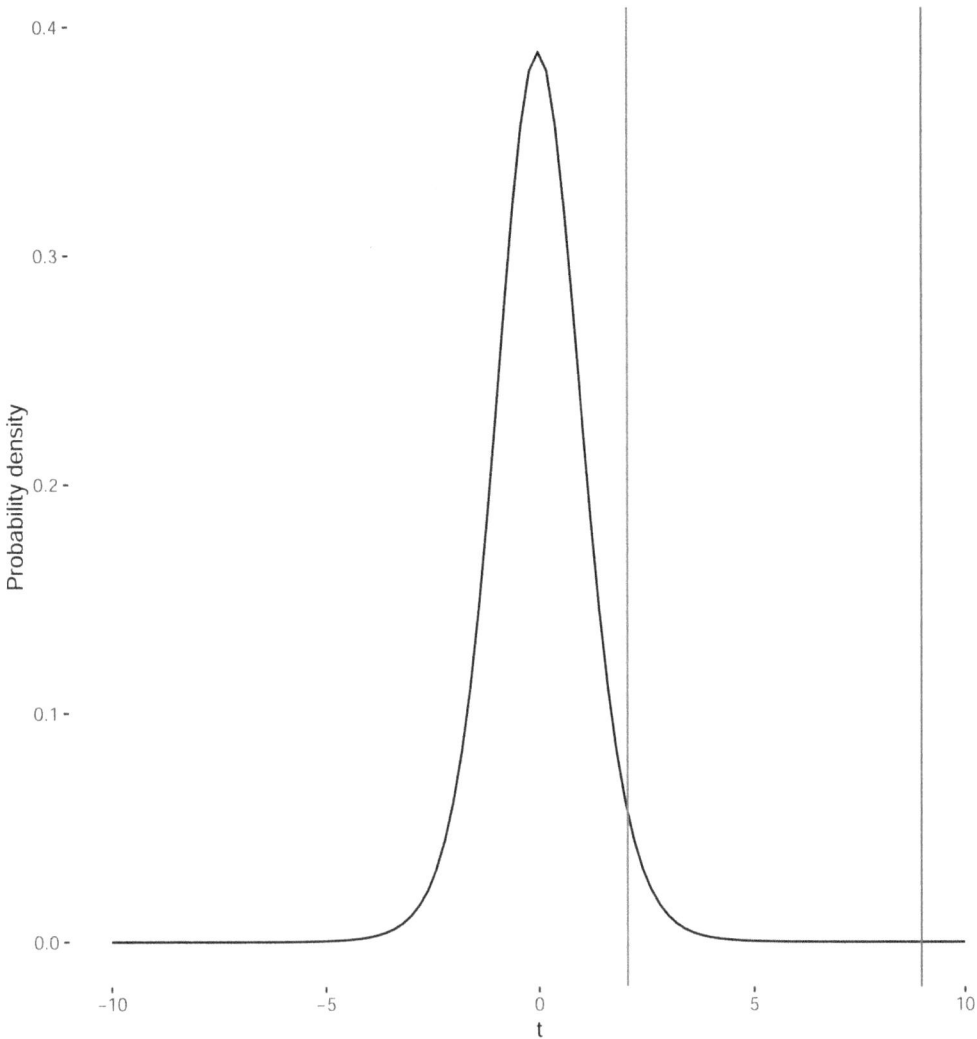

Figure 2.2 Distribution of the t statistic for 10 degrees of freedom. The blue and red lines show the value of t for the genes *ifi27.1* and *agxta*, respectively. While the value for *ifi27.1* is well within the range of reasonably probable values of t, that for *agxta* is very unlikely if the null hypothesis of no real difference in expression is true.

    b. that the gene expression values in each age group are normally distributed; and

    c. that the variance of gene expression is the same for the two age groups samples[3];

then t follows the *Student distribution* with $2n - 2 = 10$ degrees of freedom. This distribution is shown in Figure 2.2, where the values of t for *agxta* and *ifi27.1* are shown, respectively, in red and blue.

---

   [3]This assumption is easy to dispense with by using an unequal-variance t-test, but we will keep it because it is easier to generalize to a regression approach, as shown in section 2.3.

The t-value of *agxta* is in a region where the probability distribution of t in the null hypothesis is very close to zero, meaning that *if the null hypothesis were true, this value of t would be extremely unlikely*. Therefore, we can conclude with some confidence that the null hypothesis is false for *agxta*, and therefore *agxta* is indeed differentially expressed. On the other hand, the t-value of *ifi27.1* is well within the range of t-values that are likely to be obtained if the null hypothesis is true, so we have no reason to conclude that it is differentially expressed: given the high variability of its expression, the difference that we observe can very well be produced simply by random fluctuations of the measurements[4].

These intuitive considerations become quantitative with the introduction of the *P-value*:

> **i** ***P*-value**
>
> The P-value is the probability, under the null hypothesis, to obtain a value of the statistic as or more extreme than the one obtained from the data.

Thus in our case the P-value of a probe is the probability, under the Student distribution, of obtaining a value of t greater in absolute value that the one obtained from the data[5]. The P-value of *ifi27.1* is therefore the shaded area in Figure 2.3, and is equal to 0.0677: random fluctuations will produce a t as extreme as this one about 6.77% of the times. For *agxta* the P-value is much smaller, namely, $4.25 \cdot 10^{-6}$.

2.2.2 Multiple Testing

At this point we have in the P-value a satisfactory measure of how confident we are that each gene is differentially expressed, so we can simply compute the P-value for all genes, decide what P values we are willing to consider sufficiently small, and produce our list of differentially expressed genes. It turns out that the choice of the P-value threshold is not a trivial matter, and depends on the number of genes tested.

Suppose we choose a P-value threshold P_{max}, for example, the commonly used $P_{max} = 0.05$, so that we declare as differentially expressed all the genes with $P < P_{max}$. For our data, with this choice, we would declare 1,649 probes as differentially expressed and the remaining 13,968 as not differentially expressed. Inevitably, some of these calls will be wrong:

- Suppose gene G is, in reality, not differentially expressed: By definition, there is a probability of 0.05 that, just by chance, the P-value of G will be < 0.05, and thus that we will wrongly call the gene as differentially expressed. This is called a *type-I error*, or, more expressively, a *false positive*.

---

[4]Nor can we conclude that ifi27.1 is *not* differentially expressed: All we can say is that our data do not allow us to reject the null hypothesis of no differential expression

[5]More precisely one should say "greater than or equal to", but as long as the statistic has a continuous distribution this does not make a difference

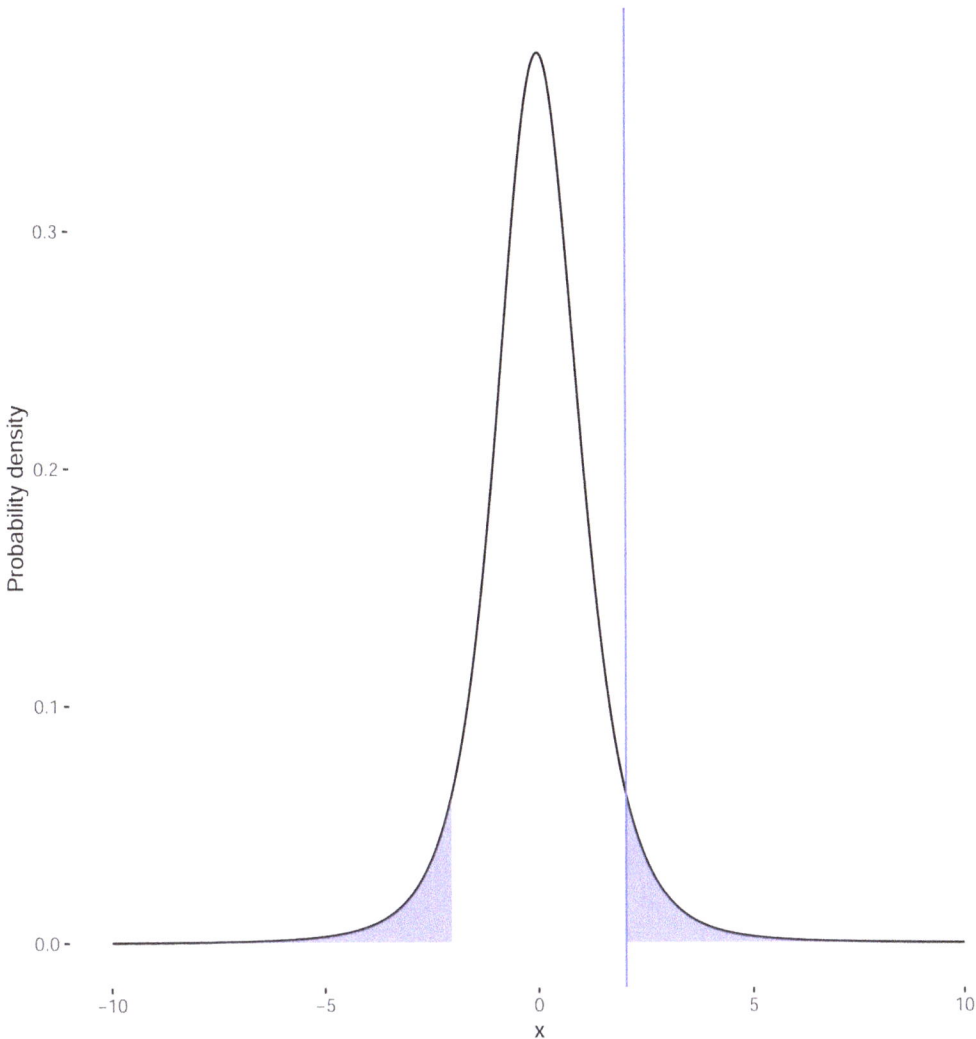

Figure 2.3 The P-value corresponding to a given value of the t statistic is computed as the probability, under the null hypothesis, to have a t-value equal or greater, in absolute value, to that obtained from the data, that is, for gene *ifi27.1*, the shaded area, which is equal to 0.06768.

- Conversely, a gene that is actually differentially expressed but whose P-value is >0.05 will not be recognized as differentially expressed, generating a *type-II error*, or a *false negative*.

Both types of errors are inevitable. Intuitively, by choosing a smaller P_{max} we reduce the false positives and increase the false negatives, and vice versa. It can be argued that false positives are a more serious type of error (they correspond to making a discovery that is in fact not true) than false negatives (failing to make a discovery that would have been true). Luckily, while we cannot eliminate false positives, we can at least *control* them, that is, estimate their prevalence and set our P_{max} in such a way as to keep their number under control.

Let us keep, for the time being, our P-value threshold at $P_{max} = 0.05$ and *estimate* the number of false positives: To do so, first assume that, in reality, no gene is differentially expressed. Then the null hypothesis is true for all genes, and we expect that $1/20$ ($= 0.05$) of them will be (wrongly) called differentially expressed. In our case, we expect $15{,}617 \cdot 0.05 = 781$ false positives. More generally, for any value P_{max} of the threshold, if the number of probes tested is N, we estimate the number of false positives to be $N \cdot P_{max}$. Two strategies are commonly used to choose P_{max}:

1. decide that we really do not want any false positives and set P_{max} in such a way that their expected number is a very small value α, for example, $\alpha = 0.05$. To achieve this we need to select $P_{max} = \alpha/N$. In our example $N = 15{,}617$ so that $P_{max} = 3.2 \cdot 10^{-6}$

2. alternatively, accept the fact that we will produce a list of differentially expressed genes that includes some false positives, but control their *rate*, that is, the ratio of the number of false positives over the total number of genes called differentially expressed. Thus, we accept to have some false discoveries, but we set a limit on the *proportion* of our discoveries that are false, the so-called *false discovery rate* (FDR). For example, in our comparison of old and young fish, it turns out that

   - for $P_{max} = 0.05$, as noted before, we make $1{,}649$ discoveries and we estimate 781 false positives, so our FDR is 0.474

   - for $P_{max} = 0.0001$ we make 24 discoveries and we estimate 1.56 false positives, so our FDR is 0.0651

In principle, we can compute the FDR for every possible value of P_{max} and choose the maximum P_{max} compatible with what we consider an acceptable FDR.

A bit more formally, the problem of multiple testing is treated by computing an *adjusted* P-value for each probe, which takes into account the number of tests performed, and then selecting as differentially expressed the probes for which this adjusted P is less than a predefined threshold.

- The *Bonferroni correction* consists in defining the adjusted P-value of a gene as:
$$P_{bonf} = min(N \cdot P, 1)$$

  where P is the original (also called *nominal*) P-value of the probe and N is the number of probes tested. This procedure follows the concept (1) above: It can be shown rigorously that if we keep only the probes with $P_{bonf} < \alpha$, the probability that one or more of these probes is a false positive is $< \alpha$.

- The *Benjamini-Hochberg procedure* follows concept (2). A (quite simple) algorithm allows us to associate to each probe, with nominal P-value P, an adjusted

P-value P_{BH} representing the proportion of false discoveries that we would expect if we were to consider as differentially expressed all the probes with P-value $\leq P$. This adjusted P-value is thus the FDR.

As an example, these are the top 10 probes by P-value, together with their log fold change, the associated gene, and the adjusted P values computed with the two procedures:

probe_id	Symbol	logFC	P	P_{bonf}	P_{BH}
Dr.25324.1.A1_at	igf2bp3	−1.14	3.793e-06	0.05923	0.01812
Dr.16410.1.A1_at	agxta	0.9406	4.251e-06	0.06639	0.01812
Dr.24852.1.S1_at	slbp	−0.4023	4.522e-06	0.07062	0.01812
Dr.12541.1.A1_at	srbd1	1.151	4.641e-06	0.07248	0.01812
Dr.16762.1.S1_at	hsbp1a	−0.5545	2.024e-05	0.316	0.0538
Dr.7946.2.A1_a_at	sh3bgrl	−0.7832	2.415e-05	0.3772	0.0538
Dr.25695.1.A1_x_at	wu:fc16g06	0.2605	2.716e-05	0.4241	0.0538
Dr.1050.1.S1_at	cdca7a	−0.6163	2.756e-05	0.4304	0.0538
Dr.4942.1.A1_at	ing4	−0.3266	3.698e-05	0.5775	0.05481
Dr.1691.2.S1_at	igf2bp1	−1.029	3.852e-05	0.6016	0.05481

Therefore, it is not possible to identify a list of differentially expressed probes by using the Bonferroni correction and $\alpha = 0.05$. However, we can identify a list of four differentially expressed probes with a FDR $< 5\%$. By going further down the list it turns out that, if we are willing to accept a FDR up to 10%, we can identify 34 differentially expressed probes.

In principle, the Bonferroni correction is preferable to the FDR, since it is obviously better to have no false positives at all[6]. In theory, when no or few discoveries survive the Bonferroni correction, one could try to improve the results by increasing the sample size (number of replicates): In general, genes that are truly differentially expressed will give smaller P values when the sample size is increased (i.e., the *statistical power* to detect differences in expression will increase), so that the number of significant P values surviving the Bonferroni correction is likely to increase. However, since transcriptomic assays are costly in terms of time and money, increasing the sample size is rarely feasible, and sometimes is simply impossible (think of studying gene expression in a rare human disease). Therefore, in many cases using the FDR instead of the Bonferroni correction is a reasonable choice.

The list of differentially expressed genes obtained by controlling multiple testing by the FDR must be used while keeping in mind that some of the genes are likely to be false positives. Thus, for example, follow-up experiments based on a specific gene in the list require its differential expression to be validated in an independent experiment. However, much useful information can be derived from such a list, for example, about the biological processes that are differentially regulated between the

---

[6]Or, more precisely, to have a very small probability of having any false positives.

two conditions (see Chapter 3), and this type of analysis is unlikely to suffer too much from the presence of a limited fraction of false positives.

Both the issue of multiple testing and the procedures used to control it are completely general, and not limited to class comparison in transcriptomics. No matter what assay we are considering, or what statistical test we use, every time we perform multiple tests we need to adjust the P values with one of the procedures described.

2.3 DIFFERENTIAL EXPRESSION AS A REGRESSION PROBLEM

The problem of finding differentially expressed genes can be alternatively formulated as a regression problem. In its simplest formulation, this is equivalent to the hypothesis testing procedure described above, but it can be generalized to take into account other variables, besides the one of interest, that could in principle influence gene expression. In our example, saying that a gene is differentially expressed is the same as saying that its expression *depends* on the age group.

> **i Regression**
>
> The term *regression* denotes a group of statistical methods for the study of the dependence among variables.

Given two variables x and y (respectively the *independent* and *dependent* variables) regression methods assume the existence of a *regression function* $f(x)$ which represents the true dependence of y on x as:

$$y = f(x) + \epsilon$$

where the *error term* ϵ represents random fluctuations, or the effects of variables that we do not consider. Various methods can then be used to estimate the regression function $f(x)$ from a series of N observations in which both x and y are measured.

The simplest methods is *linear regression*, which assumes that y depends linearly on x:

$$f(x) = \beta_0 + \beta_x x$$

The values of β_0 and β_x can be estimated from the N observations by choosing the values that minimize the *mean squared error*:

$$MSE = \frac{1}{N} \sum_{i=1}^{N} (y_i - \beta_0 - \beta_x x_i)^2$$

where x_i and y_i are the N observed values of x and y.

Returning to our experiment, let us code the age group as a numerical variable a taking values 0 for young animals and 1 for old animals[7]. Given a probe, our model is that the (logarithmic) expression y depends on the age as:

$$y = \beta_0 + \beta_a a + \epsilon$$

---

[7]The numerical values assigned to the two groups are arbitrary.

Since a takes only two values, the assumption that the dependence of y on a is linear is not really limiting.

Whether the probe is differentially expressed or not boils down to the question whether the true value of β_a is different from 0. Indeed, the true value of β_a is the difference between the expression of the probe in old vs young animals. Therefore, we can restate the null hypothesis that the probe is not differentially expressed as:

$$H_0 : \quad \beta_a = 0$$

Separately for each gene, let us estimate β_a by minimizing the MSE using our 12 observations, y_i ($i = 1 \dots 12$), six corresponding to $a = 0$ and six to $a = 1$, and call $\hat{\beta}_a$ the estimate obtained. The estimation procedure allows us also to determine the standard error of such estimate, which can be used in turn to compute a P-value[8]. The P-value represents, similarly to the case of the t-test, the probability of obtaining an estimate greater (in absolute value) than $\hat{\beta}_a$ if the true β_a were equal to zero, that is, in the null hypothesis. These are the values of $\hat{\beta}_a$ and the corresponding P values for the top 10 probes, sorted by P-value, and corrected for multiple testing with the Bonferroni and Benjamini-Hochberg procedures:

probe_id	Symbol	$\hat{\beta}_a$	P	P_{bonf}	P_{BH}
Dr.25324.1.A1_at	igf2bp3	−1.14	3.793e-06	0.05923	0.01812
Dr.16410.1.A1_at	agxta	0.9406	4.251e-06	0.06639	0.01812
Dr.24852.1.S1_at	slbp	−0.4023	4.522e-06	0.07062	0.01812
Dr.12541.1.A1_at	srbd1	1.151	4.641e-06	0.07248	0.01812
Dr.16762.1.S1_at	hsbp1a	−0.5545	2.024e-05	0.316	0.0538
Dr.7946.2.A1_a_at	sh3bgrl	−0.7832	2.415e-05	0.3772	0.0538
Dr.25695.1.A1_x_at	wu:fc16g06	0.2605	2.716e-05	0.4241	0.0538
Dr.1050.1.S1_at	cdca7a	−0.6163	2.756e-05	0.4304	0.0538
Dr.4942.1.A1_at	ing4	−0.3267	3.698e-05	0.5775	0.05481
Dr.1691.2.S1_at	igf2bp1	−1.029	3.852e-05	0.6016	0.05481

We note that these are *exactly* the same results obtained with the t-test: For each probe, the $\hat{\beta}_a$ estimate is equal to the logFC, and the P-value is the same that we obtained previously. This is both reassuring (after all we were asking the same question) and disappointing, since it seems that interpreting the problem of differential expression as a regression problem is simply a rephrasing of the hypothesis testing approach.

2.3.1 Multiple Regression

However, the regression-based approach can be easily generalized to more complex situations, in particular, to those in which the dependent variable of interest (gene

---

[8]The evaluation of the standard error and of the P-value rely on the additional assumption that the error term ϵ is normally distributed with mean 0 and variance that does depend on a: Note that these are the same assumptions that we had to make to use the t-test.

expression) depends on more than one independent variable. For example, in the dataset on the zebrafish brain, we have both male and female animals, and it is natural to expect differences in expression between the two sexes. So we can build a model in which the expression of a gene in an animal is determined by both age and sex. In the context of linear regression, the simplest such model can be written as:

$$y = \beta_0 + \beta_a a + \beta_s s + \epsilon$$

where s is 0 for females and 1 for males (again, this encoding is arbitrary). Note that this model assumes that the contributions of age and sex to gene expression are *additive* and *independent* (i.e., the contribution of age is the same for males and females, and that of sex is the same for young and old animals). We will briefly discuss more complex models below.

Linear regression is again performed by minimizing the MSE, which is now

$$MSE = \frac{1}{N} \sum_{i=1}^{N} (y_i - \beta_0 - \beta_a a_i - \beta_s s_i)^2$$

and gives us the estimates $\hat{\beta}_0$, $\hat{\beta}_a$, and $\hat{\beta}_s$, each accompanied by a P-value. The interpretation of these estimates and their P values is a bit more sophisticated: For example, $\hat{\beta}_a$ is our estimate of the difference in expression between old and young animals *once the effect of sex on gene expression has been taken into account*, and the P-value tests the hypothesis that such effect is in reality equal to zero; and similarly for $\hat{\beta}_s$.

When we fit this model, the top 10 probes sorted by the P-value associated to $\hat{\beta}_a$, together with the corresponding adjusted P values, are shown in the table below:

probe_id	Symbol	$\hat{\beta}_a$	P_a	$P_{a,bonf}$	$P_{a,BH}$
Dr.25324.1.A1_at	igf2bp3	−1.14	2.199e-06	0.03434	0.03434
Dr.16410.1.A1_at	agxta	0.9406	4.784e-06	0.0747	0.03735
Dr.12541.1.A1_at	srbd1	1.151	1.23e-05	0.1921	0.04489
Dr.24852.1.S1_at	slbp	−0.4023	1.405e-05	0.2194	0.04489
Dr.16221.1.A1_at	relt	−0.2707	1.881e-05	0.2938	0.04489
Dr.11707.2.A1_at	smox	−0.3715	1.911e-05	0.2984	0.04489
Dr.8209.1.S1_at	foxo3b	−0.1503	2.422e-05	0.3783	0.04489
Dr.10644.1.A1_s_at	fmr1	−0.4695	2.587e-05	0.404	0.04489
Dr.1050.1.S1_at	cdca7a	−0.6163	4.047e-05	0.632	0.05922
Dr.7946.2.A1_a_at	sh3bgrl	−0.7832	4.475e-05	0.6989	0.05922

While the estimates of $\hat{\beta}_a$ have not changed, the P values have. Now we can identify one probe (correspondig to gene *igf2bp3*) as differentially expressed with Bonferroni-corrected P-value less than 0.05 and eight differentially expressed genes with FDR < 5%. Therefore, keeping into account and modeling the dependence of gene expression on sex allowed us increased power in identifying genes that are differentially expressed between old and young animals.

Two remarks are necessary:

1. The fact that the estimate $\hat{\beta}_a$ does not change when adding sex to the model is somewhat peculiar of these data, and is due to the fact that sex and age are not correlated in our samples (i.e., the correlation coefficient of age and sex coded as 0 and 1 is exactly 0 for our 12 samples). In general, when the correlation between variables is non-zero also the estimate of β will change when adding a variable to the model.

2. In our case, the power to detect differential expression increased when we introduced sex in the model. This is not always the case, and the number of differentially expressed genes could very well decrease when adding more explanatory variables (called *covariates*). Nevertheless, it is always recommended to introduce as covariates all known variables that might conceivably alter gene expression.

Note also that age and sex appear in a perfectly symmetric way in our linear model: So far we were interested in the effects of age, and considered sex as a covariate, but an equally legitimate question is which genes are differentially expressed between male and female animals, while taking into account the effect of age. To answer this question, we just need to look at the estimate $\hat{\beta}_s$ for each gene and the associated P values. The top ten genes by this P-value are:

probe_id	Symbol	$\hat{\beta}_s$	P_s	$P_{s,bonf}$	$P_{s,BH}$
Dr.12369.1.A1_at	sult2st1	1.181	1.011e-05	0.1579	0.1579
Dr.14203.1.S1_at	ppp1r14aa	0.6184	3.77e-05	0.5888	0.1874
Dr.15025.1.S1_at	pdia7	0.6298	4.801e-05	0.7497	0.1874
Dr.9878.1.S1_at	atp1a3b	0.7664	7.133e-05	1	0.2228
Dr.25239.1.S1_at	sult2st2	0.6032	0.0001021	1	0.2288
Dr.8180.1.S1_at	alas2	0.7596	0.0001026	1	0.2288
Dr.12402.1.S1_at	igf1	0.9338	0.0001191	1	0.2302
Dr.19471.1.A1_at	scamp5a	0.3304	0.0001595	1	0.2302
Dr.11707.2.A1_at	smox	0.2817	0.000163	1	0.2302
Dr.4189.1.S1_at	nqo1	0.5182	0.0001707	1	0.2302

If we were studying mammals, we would find many genes differentially expressed between males and females (such as the expressed genes coded on the Y chromosome). However, this does not happen in zebrafish, in which there are no heteromorphic sex chromosomes and sex is determined by multiple genes, with some influence from the environment.

2.3.2 Limma

Limma [32] is a popular tool for regression-based differential expression analysis, allowing in particular multiple regression. While conceptually the model implemented by limma is similar to the classical regression approach described above, some improvements in the statistical treatment lead usually to increased power. These improvements are beyond the scope of these lectures; therefore, we will limit ourselves to a comparison of the results obtained by our regression approach and by limma, considering age as the independent variable of interest and sex as a covariate. These are the top ten differentially expressed probes according to limma:

probe_id	Symbol	$\hat{\beta}_a$	P_a	$P_{a,bonf}$	$P_{a,BH}$
Dr.25324.1.A1_at	igf2bp3	−1.14	8.415e-08	0.001314	0.001314
Dr.16410.1.A1_at	agxta	0.9406	3.008e-07	0.004697	0.002349
Dr.12541.1.A1_at	srbd1	1.151	5.766e-07	0.009004	0.003001
Dr.17294.1.A1_at	limch1b	−1.704	2.675e-06	0.04178	0.01045
Dr.1691.2.S1_at	igf2bp1	−1.029	3.996e-06	0.06241	0.01084
Dr.7946.2.A1_a_at	sh3bgrl	−0.7832	4.762e-06	0.07437	0.01084
Dr.13845.1.A1_at	prdx6	1.444	5.458e-06	0.08524	0.01084
Dr.1050.1.S1_at	cdca7a	−0.6163	7.744e-06	0.1209	0.01344
Dr.13614.1.S1_at	si:dkey-102m7.3	1.672	8.888e-06	0.1388	0.01388
Dr.16762.1.S1_at	hsbp1a	−0.5545	1.192e-05	0.1861	0.01692

By comparing this list with the one obtained with ordinary linear regression, we note that:

1. the top three probes are the same,
2. the beta coefficients are the same (this is true for all probes), and
3. limma P values tend to be smaller than those obtained from linear regression. Indeed, Figure 2.4 shows a plot comparing the $-\log_{10}$ of the P values for all probes.

The figure shows that most (but by no means all) probes have a smaller (more significant) P-value in limma than in ordinary regression. Therefore, the improvements to linear regression implemented by limma provided us with increased power to detect differential expression.

2.3.3 Interaction Effects

Linear regression (or limma) can also be used to ask more sophisticated questions, such as which are the genes whose expression is affected by age *differently* in male and female animals. The corresponding model is represented by the equation:

$$y = \beta_0 + \beta_a a + \beta_s s + \beta_{int} a \cdot s + \epsilon$$

The new term is called the *interaction* of age and sex, and the corresponding parameter β_{int} is different from zero if the expression y of the gene is affected by age differently in males and females (or, equivalently, if it is affected by sex differently in young and old animals). Indeed, this is the main analysis carried out in the paper

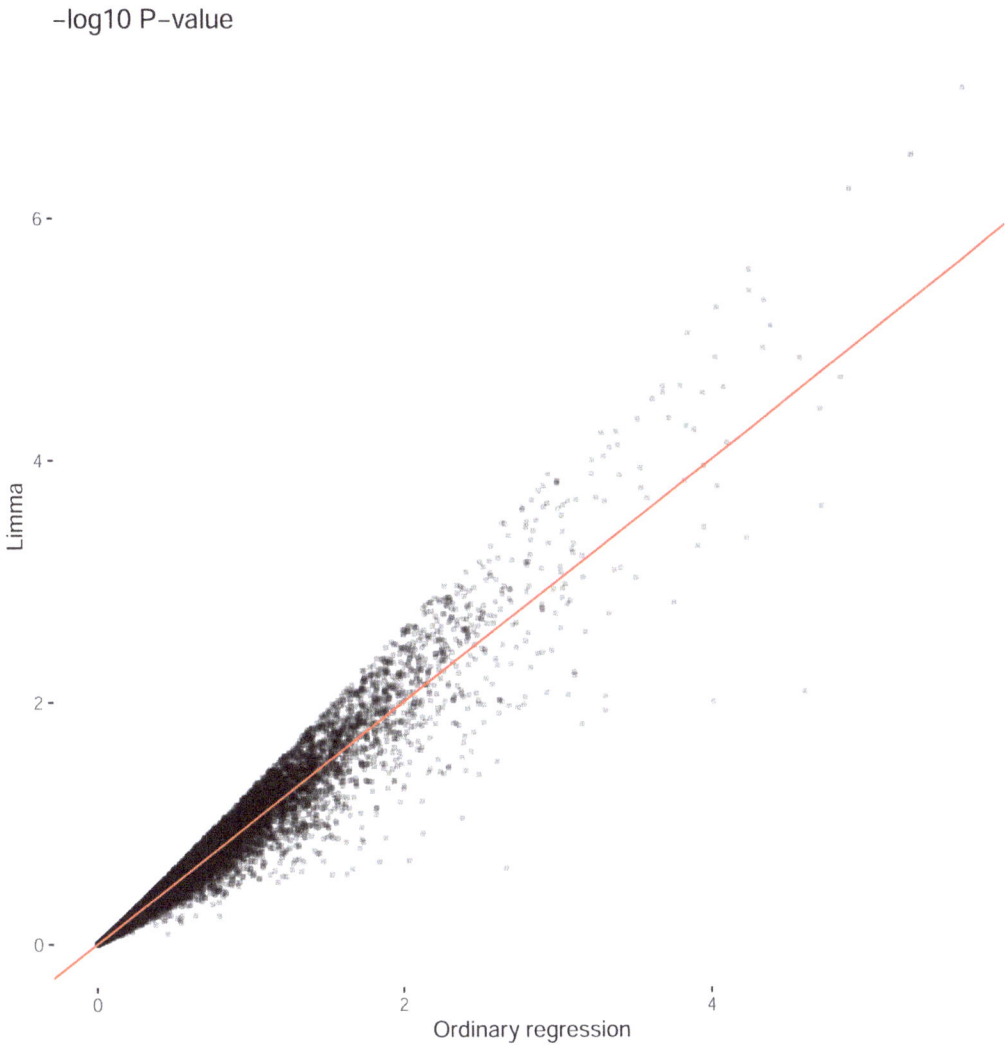

Figure 2.4 Comparison of the P values (shown as negative $\log_{10}$) found by ordinary linear regression (x-axis) and limma (y-axis) when comparing gene expression between age groups while using sex as a covariate. For the majority of the probes, limma shows increased statistical power to detect differential expression.

associated to the data [3]. For example, suppose the true expression of a gene is 0 in all animals except old males, where it is equal to 1: This situation is described by the previous equation with $\beta_0 = \beta_a = \beta_s = 0$ and $\beta_{int} = 1$, while the linear model without interaction term cannot be solved in this case[9].

---

[9]To convince yourself, write the four equations for y in each of the four possible cases (young females, young males, old females, old males) and solve for the four β values. On the other hand, you cannot solve for β in the model without interaction term.

These are the ten probes with the smallest P-value for the interaction term:

probe_id	Symbol	$\hat{\beta}_{int}$	P_{int}	$P_{int,bonf}$	$P_{int,BH}$
Dr.13782.1.S1_at	card9	−0.2826	2.346e-05	0.3664	0.3664
Dr.16186.1.A1_at	dennd4c	0.3068	0.000306	1	0.9981
Dr.7729.1.A1_at	zmynd11	0.2519	0.0003399	1	0.9981
Dr.23393.1.A1_at	wu:fl20f09	0.4088	0.0006016	1	0.9981
Dr.8138.1.S1_at	anos1b	0.5638	0.0008084	1	0.9981
Dr.15174.1.A1_at	pigo	−0.4388	0.0008358	1	0.9981
Dr.25086.1.A1_at	nr2c2	0.4123	0.001123	1	0.9981
Dr.18443.1.S1_at	zgc:56628	−0.3817	0.001322	1	0.9981
Dr.3774.1.A1_at	nek1	0.3955	0.001397	1	0.9981
Dr.728.2.A1_at	zgc:165539	−0.4537	0.002075	1	0.9981

While no probe is significant after correcting for multiple testing, the top one clearly shows the interaction effect (Figure 2.5): The expression of *card9* increases with age in females but decreases in males.

2.4 CLASS COMPARISON FOR DIGITAL EXPRESSION DATA

Several methods for measuring gene expression produce *digital* measurements, that is, they express the abundance of an RNA molecule as an integer number representing how many times a sequence related to the molecule was found among the sequences generated by the experimental apparatus. RNA-sequencing is among these methods, as are other methods that sequence only the 5′ (such as CAGE) or the 3′ (SAGE) of mRNA molecules.

The t-test we used for microarray data cannot be directly used here, since an integer variable cannot be considered as normally distributed among replicate experiments. There are two solutions to this problem: (a) Transform the data into continuous variables that can be assumed to follow the normal distribution, then use the testing/regression methods discussed above on these transformed expression data; or (b) devise testing/regression methods that can be applied directly to digital data. Route (a) is followed by limma-voom [23], where a suitable transformation into continuous expression values is followed by the application of limma. Route (b) is followed by methods such as DESeq [2] and edgeR [33]. Here we will introduce methods for direct testing/regression of count data, not because they are necessarily more effective (see e.g. [35, 6]), but because they are the most commonly used in the analysis of these data, and allow us to introduce some new concepts that will be useful also in the following chapters.

2.4.1 From Reads to Counts in RNA-seq

RNA-sequencing is a bit of a misnomer since what is actually sequenced is not RNA, but rather complementary DNA (cDNA) derived from the RNA by reverse transcription. The sequencing of the cDNA produces the same type of data as any DNA

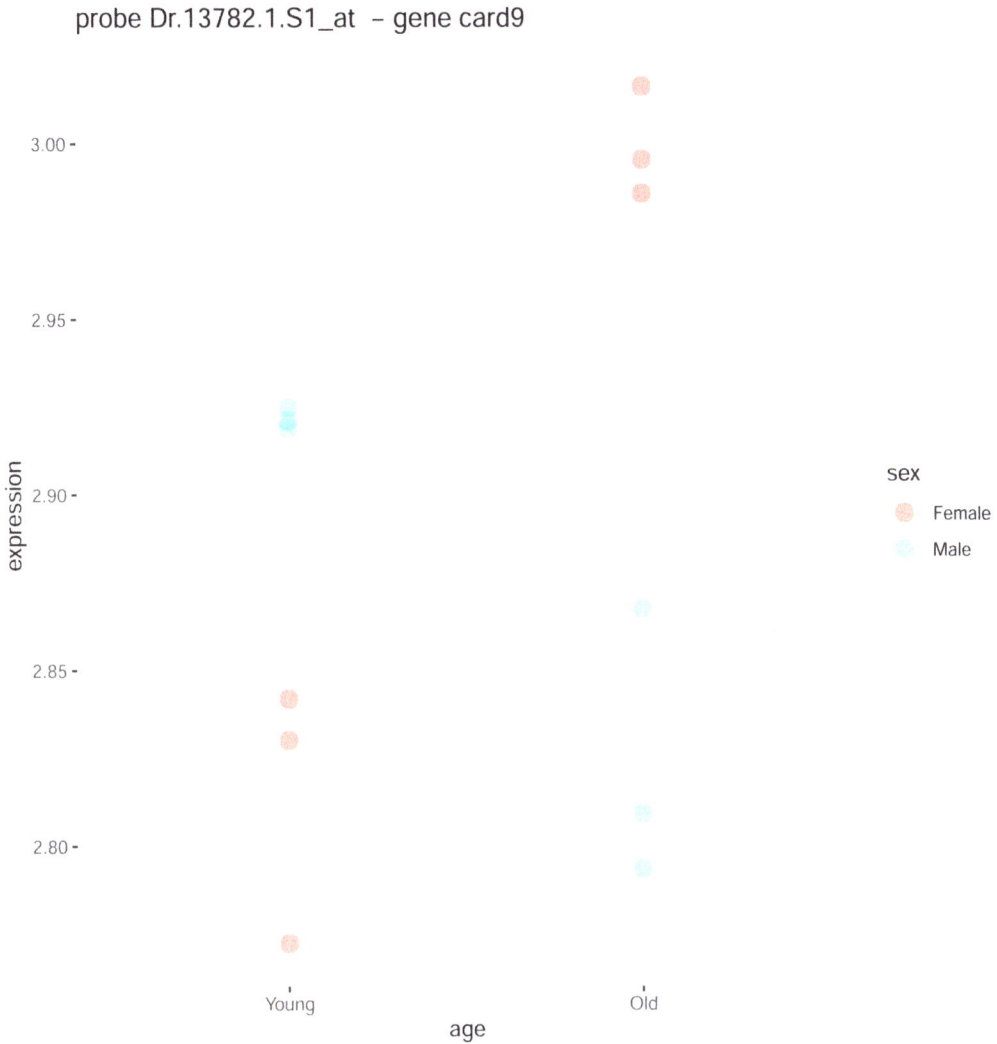

probe Dr.13782.1.S1_at – gene card9

Figure 2.5 The probe with the most significant interaction term: The expression of the corresponding gene *card9* depends on age differently for males and females.

sequencing performed with NGS, that is, a large number of short reads collected in FASTQ files. These must be transformed into *gene counts* to be useful in class comparison and other downstream analyses. A gene count is the number of reads associated to a gene, and is thus a measure of the expression of the gene, as it is proportional to the number of mRNA molecules of the gene present in the original sample[10]. How gene counts are derived from reads depends on two choices: The choice of a gene model, and the choice of a quantification strategy.

---

[10]RNA-seq also allows studying *alternative splicing*, that is, the relative abundance of the various isoforms of the same gene, a topic that we will not discuss here.

> **i Gene model**
>
> A *gene model* associated to a reference genome sequence is a catalog of all known transcripts. For each transcript, it includes:
>
> - the chromosome on which the transcript is encoded and the start and end coordinates of the transcript,
> - the strand from which the transcript is transcribed,
> - the start and end coordinates of all the exons, and
> - the start and end coordinates of the coding part, if any, of the transcript.

Gene models differ from each other based on the type of evidence required to include a transcript in the model, which translates into very different number of transcripts being included in each gene model: For example, the most recent version of the GENCODE (v. 44) gene model (which includes also transcripts supported by relatively weak evidence) catalogues $\sim$276,000 transcripts in the human genome, while the more restrictive "RefSeq curated" gene model contains only $\sim$100,000. The choice of a gene model for RNA-seq quantification depends on whether or not we are interested in quantifying also the expression of transcripts of dubious status.

After a gene model has been chosen, there are essentially two strategies to obtain gene counts from reads. In the first and more "traditional" strategy, the reads are aligned to the reference genome (with specific alignment algorithms suitable for NGS data, see Chapter 1). Then, using the gene model, we count the number of reads that were aligned to the genomic regions corresponding to the gene's transcripts[11].

The second strategy, implemented in tools such as Kallisto [4] is based on two observations:

1. If our goal is to generate gene read counts, we only care about the reads coming from known gene transcripts; reads mapping to genomic regions not associated to any known transcripts can be safely ignored.

2. we do not really care about which portion of the transcript has generated each read, but only which transcript has generated the read.

This translates into the following procedure:

1. Generate a *reference transcriptome*, based on the gene model and the reference genome, that is, a collection of the sequences of all transcripts included in the gene model.

---

[11]The alignment is complicated by the fact that the cDNA that is sequenced corresponds to the mature mRNA sequence, and thus does not include introns, which appear as large gaps when aligning the reads to the genome. Alignment algorithms used for RNA-sequencing are optimized in this sense.

2. Use *pseudoalignments* to identify the transcript, or the transcripts, that are compatible with each read. Pseudoalignment is a technique based on comparing the subsequences of length k (k-mers) contained in the read to those contained in the transcript sequence.

The main advantage of the pseudoalignment-based methods is the computational time required, which is orders of magnitude smaller that that required to align the reads to the reference genome. A possible disadvantage is the fact that this type of analysis is limited to the quantification of the expression of known transcripts, while alignment to the whole genome allows in principle the discovery of new transcripts.

Independently of which strategy was used, the reads contained in the FASTQ files are converted into a matrix of integers, with one row per gene, and one column per sample. Therefore, this matrix looks precisely like the ones we used in the analysis of microarray data, with the important difference that the numbers representing gene expression are now non-negative integers. Therefore, their statistical analysis, in particular for class comparison, must be based on a probability distribution able to model the variability of count data among replicate experiments.

2.4.2 The Poisson Distribution

The simplest choice of such a distribution is the Poisson distribution:

$$P(k; \lambda) = \frac{\lambda^k e^{-\lambda}}{k!}$$

which depends on a single parameter λ, which coincides with both the mean and the variance of the distribution. The Poisson distribution applies to many types of integer data, but is easiest to describe as modeling the number of events taking place during a specific time interval, for example, radioactive decays detected with a Geiger counter. Suppose the mean number of events recorded per minute is λ: Then $P(k; \lambda)$ gives the probability of observing exactly k events during a given one-minute interval. It follows that k is necessarily a non-negative integer, while λ is a non-negative real number.

Figure 2.6 shows a histogram of the distribution for $\lambda = 2.5$, from which we learn that if, on average, we have 2.5 events per minute, the most likely number of events observed in a given minute is two, there is a 0.082 probability of observing no events in a minute, and the probability of observing ten events is very small (actually 0.000216).

It is thus reasonable to assume that, in an RNA-seq experiment, a gene will generate reads following the Poisson distribution, with the parameter λ being proportional to the abundance of the RNA molecule (i.e., to the expression level of the gene) and to the total number of reads produced in the experiment. In the following we will consider an RNA-sequencing assay performed on two conditions, each of them assayed in some replicates. To simplify the discussion, we will also assume that the total number of reads generated in each replicate assay is the same. This is not a realistic assumption, but allows us to discuss the basic concepts behind class comparison for digital expression data while keeping the mathematical notation simple.

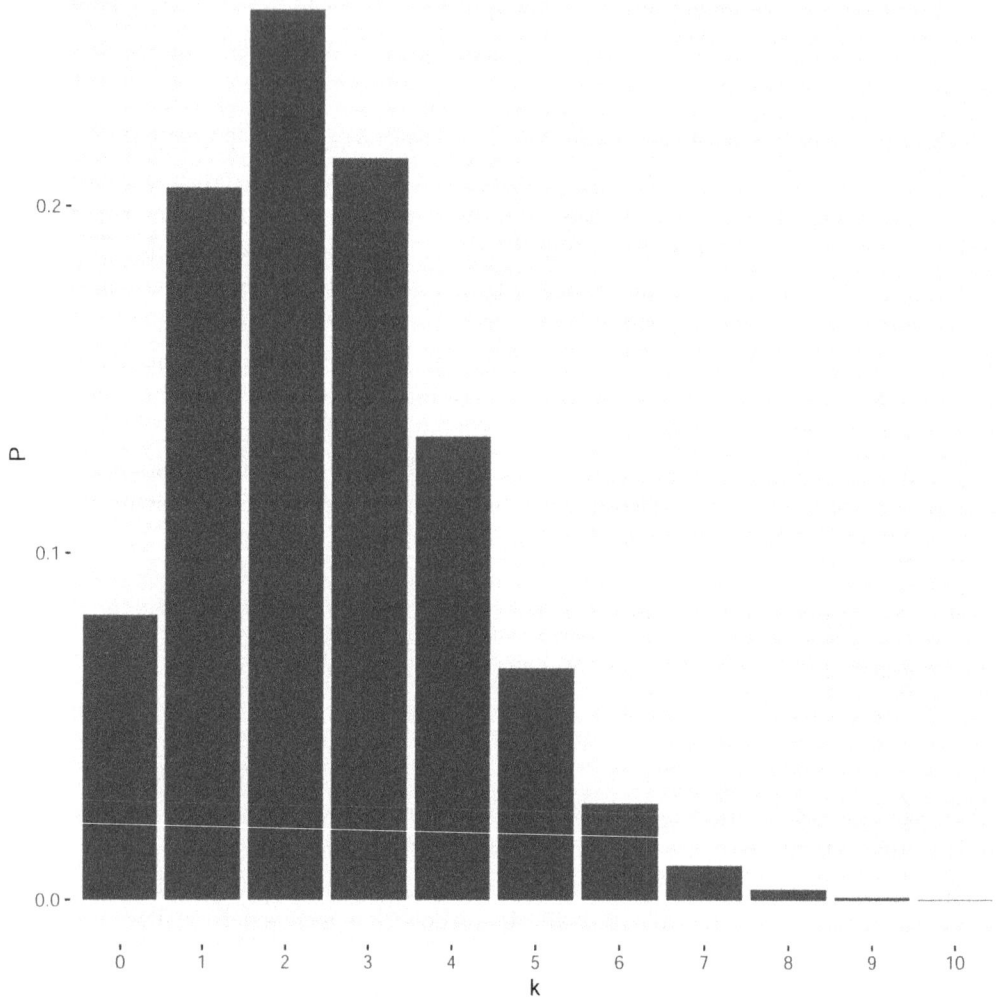

Figure 2.6 Histogram of the Poisson distribution for $\lambda = 2.5$.

Thus, we will assume that the number of reads aligned to a gene in replicate RNA-seq assays follows the Poisson distribution with mean λ proportional to the abundance of the gene mRNA in the biological condition under study. Therefore, if we have replicates from two different conditions A and B, that we want to compare, the natural null hypothesis to be tested is $\lambda_A = \lambda_B$, stating that the mRNA abundance of the gene is the same in the two conditions. This null hypothesis can be tested using a very general procedure based on the concept of *likelihood ratio*, which we will encounter again in the following chapters.

To be concrete, we will use the RNA-sequencing data produced for Ref. [28], available from the Gene Expression Omnibus under accession GSE115364, in which the transcriptome of eight samples of human astrocytes was assayed. In four of the samples the gene *IDH3A* was knocked out, while the other four samples were untreated and served as controls. Therefore, the experiment reveals the transcriptomic response of these cells to the deletion of the *IDH3A* gene: This is relevant because, as discussed in [28], *IDH3A* is an important regulator of metabolism in glioblastoma.

The statistical procedure that we will describe was introduced in [27]. After removing genes that are not expressed in either condition (10 reads or fewer when summed over all samples), we are left with a $14{,}553 \times 8$ matrix of read counts. Since we are making the simplifying assumption in which all samples produced the same number of reads, we rescale the counts to a total number of three million reads per sample. The general procedure for unequal library size does not involve any new concepts but leads to more cumbersome notation: the interested reader is referred to [27].

The counts of the first six genes (in alphabetical order) in the eight samples are shown in the table:

	CTRL_1	CTRL_2	CTRL_3	CTRL_4	KO_1	KO_2	KO_3	KO_4
A1BG	6	7	10	6	11	10	11	11
A2M	25	22	24	25	6	6	5	6
A2MP1	5	6	6	5	10	9	8	10
A4GALT	9	9	11	11	10	10	10	8
AAAS	190	192	196	203	169	187	184	165
AACS	73	72	78	73	91	98	93	90

Now, we assume that in each condition (KO or control) the normalized counts of a gene are distributed according to the Poisson distribution, with parameters λ_{KO} and λ_C for KO and controls, respectively. We want to test the null hypothesis $\lambda_{KO} = \lambda_C \equiv \lambda_0$, and, if possible, reject it to show that $\lambda_{KO} \neq \lambda_C$, that is, that the gene is differentially expressed between KO and control cells. The likelihood ratio test asks essentially whether the observed counts are better explained by a single Poisson distribution (i.e., a single value $\lambda_0 = \lambda_C = \lambda_{KO}$) or by two distributions with two different values $\lambda_{KO} \neq \lambda_C$.

2.4.3 Maximum Likelihood Estimation

> **i Likelihood**
>
> The *likelihood* of a probability distribution given some observed data is the probability to generate those data from the distribution.

For example, suppose a Geiger counter records ten decays in a minute. The likelihood of the Poisson distribution with $\lambda = 2.5$ is

$$L(\lambda = 2.5 | k = 10) = P(k = 10; \lambda = 2.5) = 0.000216$$

The likelihood is small since it is unlikely that a Poisson distribution with mean $\lambda = 2.5$ will generate ten decays in a minute. Indeed the likelihood of $\lambda = 10$ is much

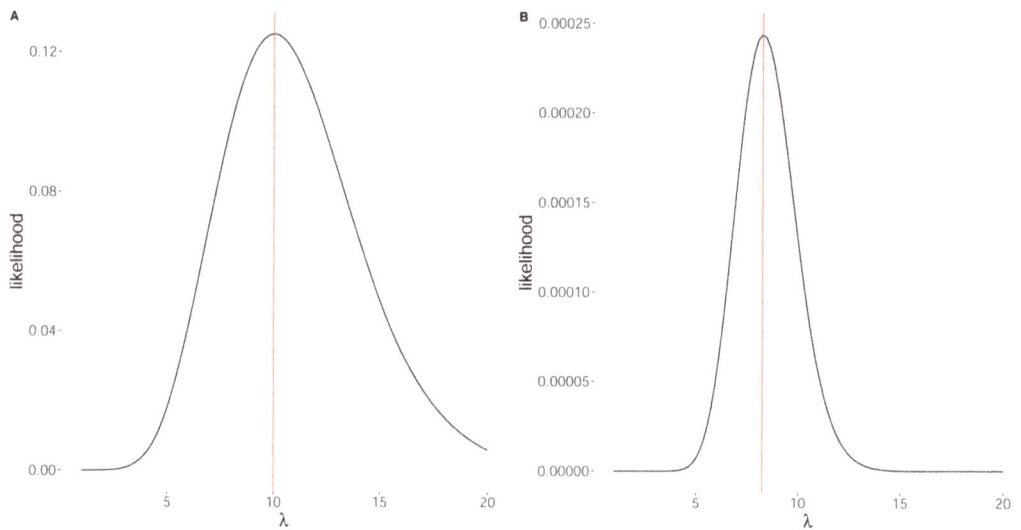

Figure 2.7 (A) Likelihood as a function of λ when $k = 10$ events were observed. (B) Same if four independent observations produced $k = 7, 9, 7,$ and 10 events.

higher:

$$L(\lambda = 10 | k = 10) = P(k = 10; \lambda = 10) = 0.125$$

This suggests a general method to estimate a distribution parameter, such as λ, from data:

> **i Maximum likelihood estimation**
>
> Given some data, we can estimate one or more parameters of a probability distribution by choosing as estimate the value of the parameters corresponding to the maximum possible likelihood. Therefore, the estimated parameters are the ones that would generate the observed data with the highest probability.

Figure 2.7A shows the likelihood as a function of λ for $k = 10$ observed decays: Unsurprisingly, the maximum likelihood is for $\lambda = 10$. Therefore, if we observe ten decays in a minute, our estimate of λ is 10. This is called the *maximum likelihood estimate* (MLE) of λ. If we have several independent measurements, we compute the likelihood as a product: For example, if we observed 7, 9, 7, and 10 decays in four different minute-long measurements, the likelihood as a function of λ is

$$L(\lambda | k = 7, 9, 7, 10) = P(k = 7; \lambda)^2 \cdot P(k = 9; \lambda) \cdot P(k = 10; \lambda)$$

shown in Figure 2.7B, and our estimate of λ is the value corresponding to the maximum likelihood, namely $\hat{\lambda} = 8.25$.

It turns out that for some (but not all) distributions, including the normal and Poisson distributions, the MLE of the distribution mean is equal to the mean value of the observed data. Clearly, this method of estimation can be generalized to any distribution, and has many other virtues that have made it one of the most popular methods of statistical analysis.

2.4.4 Likelihood Ratio Test

Returning to our null and alternative hypotheses ($\lambda_{KO} = \lambda_C$ and $\lambda_{KO} \neq \lambda_C$, respectively), the likelihood ratio test is based on computing the ratio of the maximum likelihoods of the two hypotheses given the observed data, that is:

- L_0, the maximum likelihood that can be obtained assuming that the counts from both conditions follow Poisson distributions with the same λ.

- L_1, the maximum likelihood that can be obtained assuming that the counts from each condition follow Poisson distributions with λ values λ_{KO} and λ_C that are not constrained to be equal.

It is important to note that L_1 will *always* be greater than or equal to L_0, as the model with two λ parameters contains, as a special case, the model with a single value (the two models are said to be *nested*). Even if the null hypothesis is true, we will always have a gain in likelihood when allowing the two parameters to be different.

> **i   Likelihood ratio test**
>
> The likelihood ratio test determines whether the gain in likelihood obtained by allowing the two conditions to have different λ values is statistically significant. The corresponding P-value is the probability to obtain a gain in likelihood as large or larger than the one obtained if the null hypothesis is true.

As usual, a small P-value will indicate that the null hypothesis is probably not true, and hence that the gene is differentially expressed between the two conditions. The test is made possible by the fact that the statistic:

$$-2\ln(L_0/L_1)$$

follows the χ^2 distribution with one degree of freedom if the null hypothesis is true.

Let us apply this test to the genes *A1BG* and *A2M* whose read counts were shown in the table above and are represented graphically in the barplots in Figure 2.8. For *A1BG*, the mean expression over all samples (which coincides with the MLE of λ in the null hypothesis) is $\hat{\lambda}_0 = 9$, while the ones computed on controls and KO are, respectively, $\hat{\lambda}_C = 7.25$ and $\hat{\lambda}_{KO} = 10.8$. We can then compute:

$$\ln(L_0) = \sum_{i=1}^{4} \ln P(n_i^C; \hat{\lambda}_0) + \sum_{i=1}^{4} \ln P(n_i^{KO}; \hat{\lambda}_0) = -18.2$$

where n_i^{KO} and n_i^C are the counts for KO and control, while

$$\ln(L_1) = \sum_{i=1}^{4} \ln P(n_i^{KO}; \hat{\lambda}_{KO}) + \sum_{i=1}^{4} \ln P(n_i^C; \hat{\lambda}_C) = -16.83$$

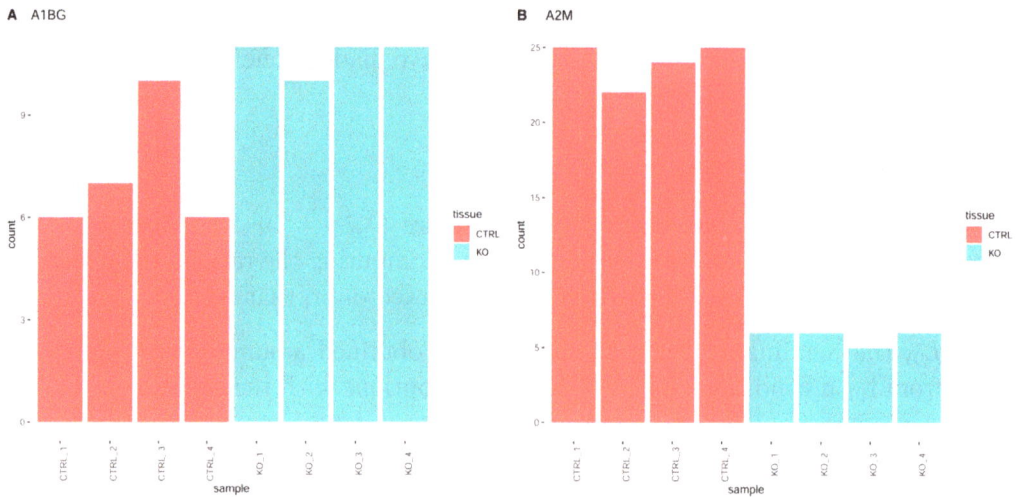

Figure 2.8 Barplot of the counts for genes *A1BG* (A) and *A2M* (B) in KO and wild-type replicates.

and

$$\chi^2 = -2\ln(L_0/L_1) = 2.74$$

The same procedure applied to *A2M* gives $\chi^2 = 48.1$.

The χ^2 distribution with one degree of freedom is shown in Figure 2.9 with the χ^2 values for *A1BG* and *A2M* shown in red and blue, respectively. The *P*-value is computed as the area to the right of the χ^2 value for each gene[12]. For *A1BG* and *A2M* we have, respectively, $P = 0.0979$ and $P = 4.0 \cdot 10^{-12}$.

Therefore, we can say that the alternative hypothesis (the two conditions are described by Poisson distributions with different λ parameters) always has higher likelihood than the null hypothesis (same λ). For *A2M*, but not for *A1BG*, the ratio between the two likelihoods is sufficiently large to be very unlikely under the null hypothesis, so that we can declare *A2M*, but not *A1BG*, to be differentially expressed. When the procedure is applied to all genes, the same multiple testing issue arises that we saw when discussing the *t*-test, and can be treated with the same tools, including the Bonferroni correction and the Benjamini-Hochberg procedure to calculate the FDR.

2.4.5   Better Statistical Models

Various improvements to this framework have been proposed and applied in the software packages commonly used for differential expression, such as DESeq2 [26] and edgeR [33]. In edgeR, for example, the following improvements are introduced with respect to the Poisson model:

---

[12]Note that since, as discussed above, we certainly have $L_0 \leq L_1$, χ^2 is always non-negative, and only large and *positive* χ^2 values suggest rejecting the null hypothesis. Hence, the *P* values are computed using only the right tail of the distribution, contrary to what we did for the Student distribution used in the *t*-test.

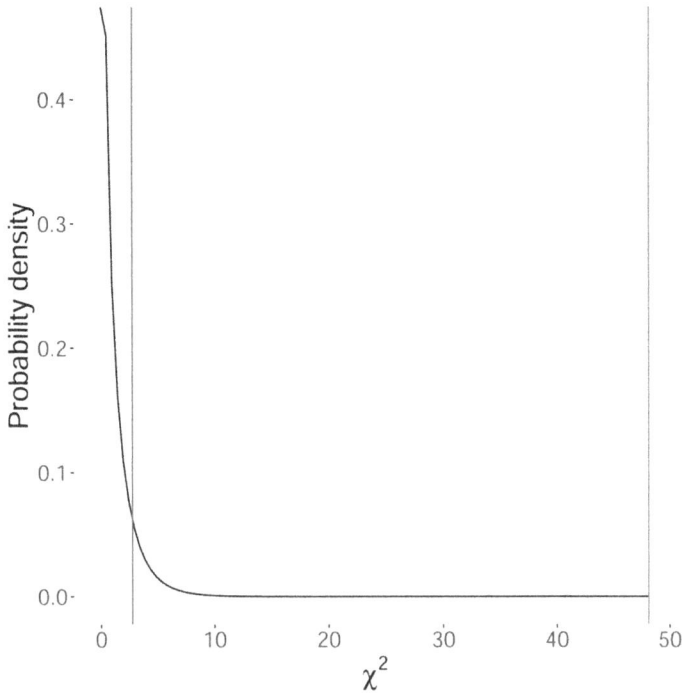

Figure 2.9 The χ^2 probability distribution with one degree of freedom, with the χ^2 values for genes *A1BG* (red) and *A2M* (blue).

- A *negative binomial* distribution is used to model the count data: Compared with the Poisson distribution, the negative binomial has an additional parameter (the dispersion) that allows the variance to be different from the mean. The Poisson distribution is recovered as the special case of zero dispersion. The negative binomial distribution has been shown to capture the inter-replicate variation better than the Poisson distribution, and using the Poisson distribution has been shown to underestimate variability, thus leading to an increase in false positives.

- For simple experimental designs (comparing two classes), differential expression can also be evaluated using an exact test similar to the Fisher test (see Chapter 3) rather than with the likelihood ratio test.

- A regression approach can be used for complex experimental designs involving several factors and their interactions. Conceptually similar to the linear regression models described above, this approach relies on generalized linear models, namely regression models in which the response does not depend linearly on the independent variables.

In Figure 2.10, we compare, for each gene, the P values given by our Poisson-based procedure and the negative binomial-based edgeR (the exact test implemented by edgeR gives very similar results). We note that there is a strong correlation, as

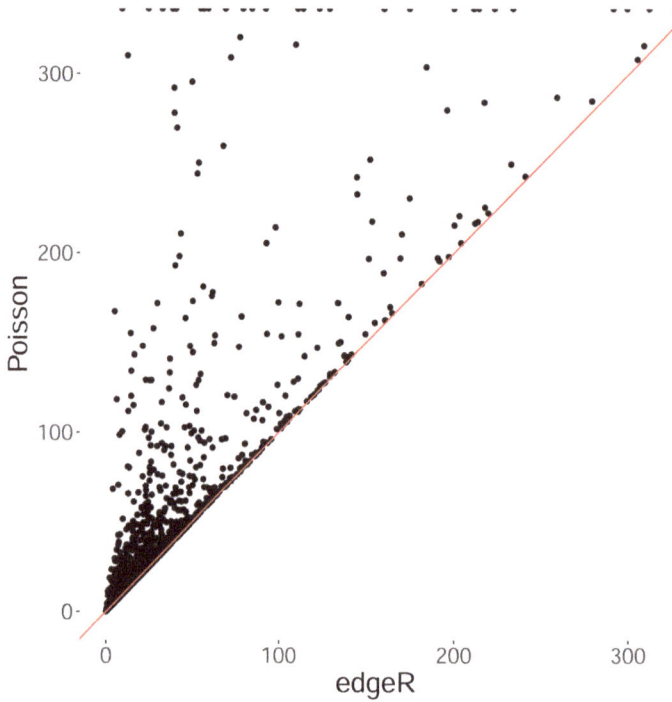

Figure 2.10 Comparison of the P values obtained from edgeR (x-axis) and the Poisson-based likelihood ratio test (y-axis) for all genes. P values are represented as their -$\log_{10}$, and the red line is the bisector $y = x$. Poisson P values are systematically more significant than the edgeR ones.

expected, between the P values. However, the Poisson P values are systematically more significant (dramatically, for some genes) than the edgeR ones. This is due to the fact that actual RNA-seq count data are better represented by the negative binomial distribution than by the Poisson distribution, and this results in an "inflation" of P values when using the Poisson distribution. More specifically, the variance of RNA-seq counts among replicates is greater than the one predicted by the Poisson distribution (*overdispersion*). This is especially true when the variance has biological rather than technical origin (e.g., the replicates are, human patients rather than batches of cell lines). In these cases it is thus especially important to use a method based on the negative binomial distribution, such as edgeR or DESEq2.

FURTHER READING

Berge, K., Hembach, K., Soneson, C., Tiberi, S., Clement, L., Love, M., Patro, R. & Robinson, M. RNA sequencing data: Hitchhiker's guide to expression analysis. *Annu Rev Biomed Data Sci.* **2**, 139–173 (2019).

Love, M., Huber, W. & Anders, S. Moderated estimation of fold change and dispersion for RNA-seq data with DESeq2. *Genome Bio.* **15**, 550 (2014).

Marioni, J., Mason, C., Mane, S., Stephens, M. & Gilad, Y. RNA-seq: An assessment of technical reproducibility and comparison with gene expression arrays. *Genome Res.* **18**, 1509–1517 (2008).

Ritchie, M., Phipson, B., Wu, D., Hu, Y., Law, C., Shi, W. & Smyth, G. limma powers differential expression analyses for RNA-sequencing and microarray studies. *Nucleic Acids Res.* **43**, e47 (2015).

Robinson, M. & Smyth, G. Small-sample estimation of negative binomial dispersion, with applications to SAGE data. *Biostatistics.* **9**, 321–332 (2008).

CHAPTER 3

Transcriptomics, Part II: Class Discovery and Enrichment Analysis

3.1 INTRODUCTION

Class discovery consists in analyzing a dataset (such as the results of a transcriptomic assay) by identifying (*discovering*) groups of data with similar features. In transcriptomics, this type of analysis is usually performed on experiments that do not simply involve replicates of a small number of conditions, but have a more complex design, such as a time course in which a dynamical process is followed by performing transcriptome measurements at several timepoints, or a collection of individual samples of a tissue, for example, the surgically resected tumors of a set of cancer patients.

As it is often the case, the algorithms used in class discovery did not originate in genomics, but are generic tools that can be used on data of different origin: The algorithms we will describe can be used, for example, to analyze financial, astronomical, or social data: whenever we need to partition a large class of objects (stocks, galaxies, users of a social network) into subgroups in such a way that the objects within a group are more *similar* (to be defined precisely) to each other than to those in other groups.

Let us consider a transcriptomic assay, for example, the expression profiles of a large set of breast tumors. There are (at least) two ways in which class discovery can be used on such a dataset:

- We can discover classes of *genes* whose expression profiles across the samples are similar to each other. This is useful because, as we shall see, often the genes within each such class are involved in similar biological processes. Therefore, the discovery of such classes can be useful in *functional genomics*, that is, the attempt to understand the biological function of all the genes in the genome.

- We can discover classes of *samples* with similar transcriptomes. These are classes of tumors that differ at the molecular level, and such molecular

DOI: 10.1201/9781003449928-3

differences could correlate with different prognosis or different responses to treatment.

Class discovery is often performed with *clustering* algorithms. These algorithms require a quantitative *dissimilarity measure* between data, that is, a rule that, given two sets of data (for example the expression profiles of two genes), produces a number expressing how similar they are to each other. We will first discuss two possible dissimilarity measures (among many that have been proposed) and then describe some clustering algorithms needed to create *clusters* of data based on such measures.

Both the dissimilarity measure and the clustering algorithm can be chosen in several ways, and we do not have (yet) a solid theoretical foundation to help us choose among them based on the nature of the data and the biological question of interest. Therefore, these choices are often performed in an empirical way, and the final test of our choices is whether the biological information produced by the analysis can be independently validated. In their pioneering study on clustering in transcriptomics of 1999, Tavazoie et al. [37] referred to their choice of dissimilarity measure as follows, and the situation has not really changed since:

> *Other metrics are also used in multivariate clustering, and our use of the Euclidean distance reflects our ignorance of a more biologically relevant measure of distance.*

3.2 CLUSTERING

3.2.1 Data as Vectors

The data that we want to cluster can be represented as *vectors*.

> **i Vectors and dimensionality**
>
> A *vector* is an ordered list of real numbers, the vector *components*. The number of components of a vector is called its *dimensionality*.

We refer to the number of components as dimensionality because an ordered list of D real numbers can be interpreted geometrically as the coordinates of a point living in a D-dimensional space. We will exploit this geometric analogy below when defining dissimilarity measures.

For example, consider a transcriptomic assay measuring the mRNA abundance of 6,000 genes in seven biological conditions:

- The *genes* are represented by 6,000 seven-dimensional vectors: Each vector is defined by seven components, each equal to the expression of the gene in one of the assayed samples.

- Similarly, the *samples* are represented by seven 6,000-dimensional vectors.

To avoid being too abstract, in the following we will introduce dissimilarity measures and clustering algorithms as applied to the genes, rather than to the samples, but all the math is exactly the same when applied to the clustering of the samples, even if the biological interpretation is completely different.

3.2.2 Dissimilarity Measures

We will work on data from a classic microarray experiment, published in 1997 by DeRisi et al. [7]. This experiment demonstrated how microarray technology could be used to study the transcriptomic changes implemented by the cell (in this case yeast) to adapt to a change in the environment. Specifically, the transcriptome of yeast cells was measured at seven timepoints during ~24 hours, while the concentration of glucose in the growth medium was progressively depleted. The cells react to this change in the environment by switching from anaerobic fermentation to aerobic respiration as the source of energy for growth. Since the cell cannot change its genome to respond to environmental changes, this metabolic shift, called the *diauxic shift*, is effected by changes in gene regulation, which can be studied by measuring the transcriptome at various timepoints during the shift.

The data are publicly available in the Gene Expression Ominbus (GEO) repository under accession GSE28. Expression was measured in 6,400 probes, 6,153 of which are associated to a yeast gene. The expression is given, as usual, in an appropriate log_2 scale (see the GEO site for the precise definition). The table below represents the measured expression of ten yeast genes (chosen for illustrative purposes) at each timepoint:

Gene	t_0_h	t_9.5_h	t_11.5_h	t_13.5_h	t_15.5_h	t_18.5_h	t_20.5_h
ACS1	−1.07	−0.514	−0.22	−0.012	−0.215	1.741	4.239
CBP6	0.006	0.136	0.399	−0.032	−0.297	1.098	1.071
COX8	0.133	0.124	0.371	0.537	1.181	1.447	2.095
FLO5	−0.109	−0.014	−0.252	−0.243	−0.008	−0.339	0.752
ICL2	−0.356	−0.37	−0.41	−0.333	0.185	1.005	1.449
MCM2	−0.266	0.269	−0.04	−0.268	−0.579	−0.304	−0.971
RPL27A	0.617	0.417	0.535	0.214	0.221	−1.033	−1.584
RPS5	0.165	0.373	0.09	−0.184	−0.173	−1.789	−2.097
TRS33	−0.027	0.104	0.131	0.313	0.653	0.189	−0.388
UBC8	−0.176	−0.108	0.105	0.253	−0.091	1.687	2.842

In Figure 3.1 each gene is represented graphically by a line with time on the x-axis and expression on the y-axis. Intuitively, the expression profiles of *ACS1*, *COX8*, and *UBC8* are similar to each other and very dissimilar from those of *MCM2*, *RPL27A*, and *RPS5*. We need to define a quantitative measure of this intuitive notion of dissimilarity.

Incidentally, just by looking at the plot we could perform our class discovery "by hand" by identifying genes with similar behavior. For example, we could define a class of genes that are upregulated at the end of the diauxic shift (*ACS1*, *UBC8*, and *COX8*), one of genes whose expression remains essentially stable (*ICL2*, *CBP6*, and *FLO5*), and one of downregulated genes (*TRS33*, *MCM2*, *RPL27A*, and *RPS5*).

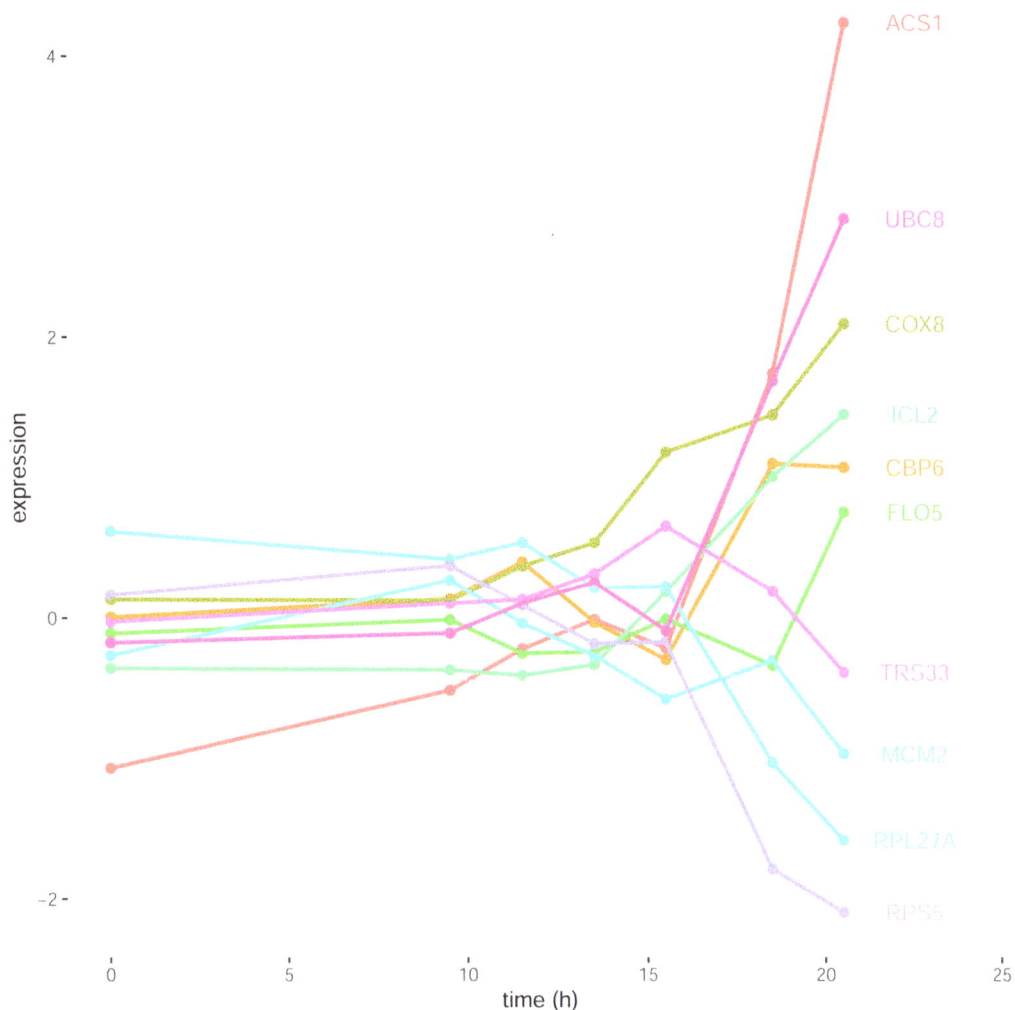

Figure 3.1 Graphical representation of the expression change in time for ten yeast genes during the diauxic shift experiment.

There are two reasons why we prefer to use a clustering algorithm instead of proceeding by hand. The first reason is trivial, and is the fact that in real-case scenarios we are dealing with thousands of genes, and not just ten. However, it would not be especially difficult to develop software in which we create our clusters using the mouse from a graphical representation of the expression profiles similar to the one shown above.

The second reason is more subtle, and has to do with the unavoidable biases that human investigators have toward their research. When we work on a research project we are rarely neutral with respect to the results of the experiments (real or computational). Therefore, if we perform the analysis by hand, we run the risk of being influenced by our preconceptions. In our example, the gene *TRS33* could reasonably be associated to the "down" or the "stable" class, and we might be biased toward one of the two choices because it agrees better with what we are trying to show. Then we could, consciously or not, choose the partition into classes that better agrees with our theory; or choose the other partition in an effort to avoid biasing the analysis. To avoid these dilemmas, we simply trust a computer to do the job using an algorithm, knowing that neither the computer nor the algorithm have any emotional involvement in our results.

3.2.3 Euclidean Distance

Probably the simplest quantitative measure of dissimilarity is the Euclidean distance. Being simple to understand and to compute, it is a good choice when we lack the theoretical basis to make a more informed one, as noted in the quote above from [37]. A D-dimensional vector (in our case a gene is a vector with $D = 7$) can be represented as a point in a D-dimensional Euclidean space by interpreting each component as a coordinate on a Cartesian axis.

> **i Euclidean distance**
>
> In a transcriptomic assay measuring gene expression in D conditions, the *Euclidean distance* between two genes is the geometric distance between the corresponding points in the D-dimensional space.

Let us temporarily use only the last two timepoints in the diauxic shift (18.5 and 20.5 hours), so that $D = 2$ and each gene can be represented as a point in a plane, as shown in Figure 3.2[1]. Then a natural definition of distance between two genes is the length of the straight line segment between them: For example, the distance between *ACS1* and *RPL27A* is 6.45 and that between *ACS1* and *COX8* is 2.16, that is, the lengths of the segments represented in the figure.

The Euclidean distance can be readily generalized to two vectors x and y with an arbitrary number D of components:

$$d(x, y) = \sqrt{\sum_{i=1}^{D} (x_i - y_i)^2}$$

where $x_1 \ldots x_D$ are the D components of vector x. This allows us to compute the Euclidean distances between our genes taking into account all the seven timepoints. When we do so, we get, for example, the following distances between *ACS1*, *COX8*, and *RPL27A*.

	ACS1	COX8	RPL27A
ACS1	0	3.023	6.792
COX8	3.023	0	4.589
RPL27A	6.792	4.589	0

---

[1]Note that this graphical representation is different from the one used in Figure 3.1.

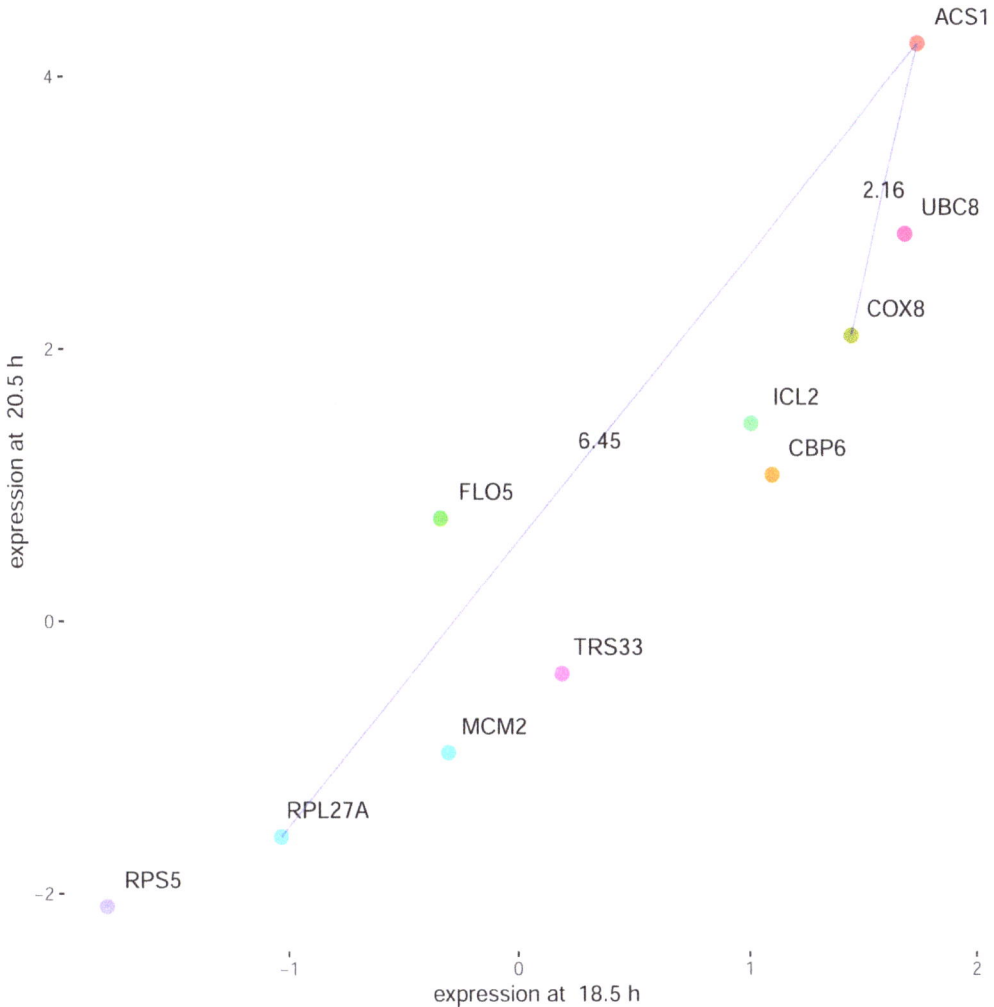

Figure 3.2 Genes represented as points in a plane. The coordinates of each gene are its expression levels in the last two timepoints of the diauxic shift experiment. The Euclidean distance between two genes is the length of the straight line segment joining the two genes.

The number themselves are not especially meaningful, but it is important to note that our intuitive notion of dissimilarity in expression profiles is captured by this quantitative measure: *ACS1* is closer to *COX8* than to *RPL27A*, in agreement with our intuition when looking at the plot of their expression as a function of time (Figure 3.1).

3.2.4 Correlation-Based Dissimilarity

Another measure of dissimilarity often used in clustering gene expression data is based on the *Pearson correlation coefficient* between vectors, defined as:

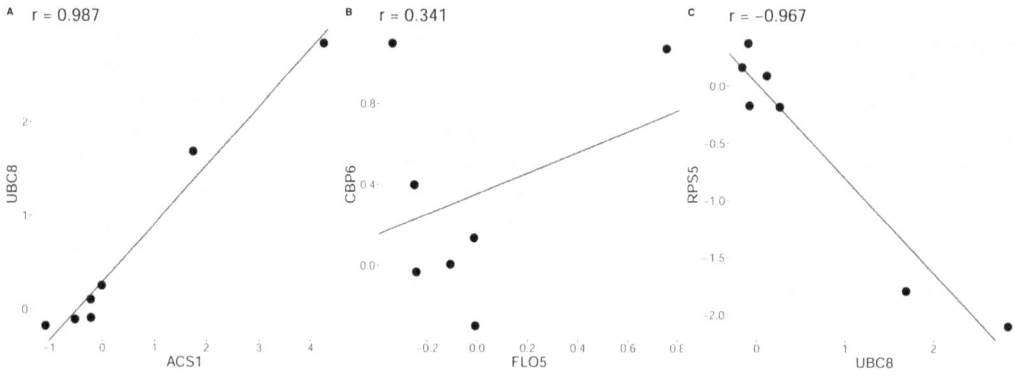

Figure 3.3 Examples of correlation between genes, used to compute the correlation-based dissimilarity. Note that in this graphical representation, which differs from the ones used before, the x and y axes represent two genes, and each dot one timepoint. The coordinates of the dot are the expression values of the two genes at that time-point. The straight line represents the best linear regression. (A) Genes $ACS1$ and $UBC8$ show strong positive correlation. (B) $FLO5$ and $CBP6$ show weak positive correlation. (C) $UBC8$ and $RPS5$ have strong negative correlation.

$$r(x, y) = \frac{\sum_{i=1}^{N}(x_i - \bar{x})(y_i - \bar{y})}{\sqrt{s_x^2 s_y^2}}$$

where $\bar{x}$ is the mean and s_x the standard deviation of x. It can be shown that $-1 \leq r \leq 1$, and that r is a measure of the linear dependence between x and y. Values $r > 0$ ($r < 0$) indicate positive (negative) correlation, that is that the best straight line fit of y as dependent on x has positive (negative) slope, while the absolute value of r indicates how well a straight line fits the dependence of y on x. This should remind you of linear regression, introduced in Chapter 2, and indeed there is a close relationship between r^2 and the mean squared error of the linear regression of y as dependent on x (or of x as dependent on y).

Going back to the genes we used as examples, $ACS1$ and $UBC8$ show strong positive correlation (Figure 3.3A - the straight line represents the best linear regression); $FLO5$ and $CBP6$ show an example of weak positive correlation (Figure 3.3B), while $UBC8$ and $RPS5$ have strong negative correlation (Figure 3.3C).

Since we need a *dissimilarity* measure that, like the Euclidean distance, grows larger for genes whose expression profiles are more dissimilar, we use as dissimilarity a decreasing function of r, such as:

$$d_{corr}(x, y) = \frac{1 - r(x, y)}{2}$$

which is 0 for two perfectly and positively correlated genes, and reaches its maximum value of 1 for perfectly anticorrelated genes ($r = -1$).

Compared with the Euclidean distance, the correlation-based dissimilarity considers as similar two genes showing the same behavior irrespective of their actual expression values, so that if the expression of gene x is, in each sample, exactly ten times that of gene y, their correlation dissimilarity will be zero, even if their Euclidean distance is far from being zero. Thus, one could argue that correlation-based dissimilarity should be used when the shape of the expression profile is biologically more significant than the expression values themselves. However, in practice, it is difficult to judge *a priori* when this is the case, and the choices of dissimilarity measures are often compared *a posteriori* based on their agreement with previous biological knowledge, for example, in terms of the guilt by association principle discussed below (see, e.g. [19]).

3.2.5 Clustering Algorithms

Given a quantitative measure of dissimilarity between vectors, we can proceed to classify our genes into clusters using one of several available *clustering algorithms*.

> ⓘ **Clustering**
>
> Clustering algorithms partition a set of vectors into *clusters* in such a way that:
>
> - vectors within a cluster are more similar to each other than to vectors belonging to other clusters;
> - each vector belongs to one and only one cluster.

Some clustering algorithms remove the second requirement and allow each vector to belong to different clusters with different probabilities. This is called *fuzzy clustering*.

Many clustering algorithms have been proposed, which can be classified based on the general principles they adopt to perform the job. Two of the most commonly used classes of algorithms are hierarchical clustering and centroid-based clustering. We will describe in some detail an example of each of these two classes, while remembering that, as discussed above, the choice among different algorithms is mostly dictated by empirical arguments due to the lack of a solid theoretical structure that could guide the choice. Network-based clustering is a third approach that is often used in the analysis of single-cell expression data, and will be discussed in Chapter 4.

3.2.6 Average-Linkage Hierarchical Clustering (UPGMA)

> ⓘ **Hierarchical clustering**
>
> In *hierarchical clustering*, a tree is built using a method from phylogenetic analysis, such as UPGMA, with the genes to be clustered as the leaves, and the dissimilarity measure used as distance. Then the tree is cut at a given value of the dissimilarity measure, and each branch generates a cluster containing all its descendant genes.

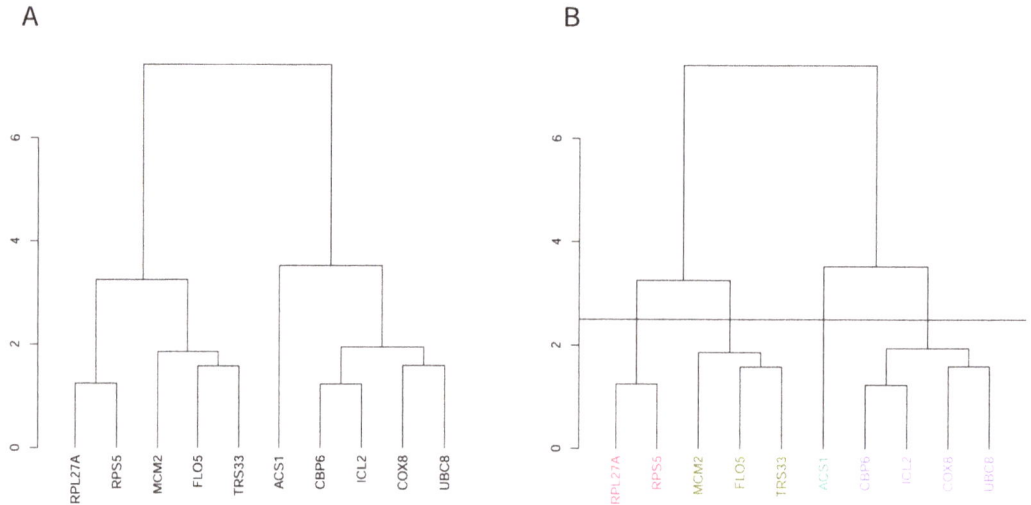

Figure 3.4 A: Hierarchical clustering creates a 'phylogenetic tree' with the genes to be clustered as the leaves, based on the dissimilarity measure chosen. B: To find the actual clusters, the tree is cut at a given height (value of the dissimilarity measure), and each branch corresponds to a cluster containing all its "descendants."

Note that the distance and the tree do *not* have an evolutionary meaning, and should be considered as mere mathematical tools used to build the clusters.

For example, the tree obtained using UPGMA and the Euclidean distances among our ten genes (again using all seven timepoints) is shown in Figure 3.4A. The actual clusters are obtained by cutting the tree at a certain height h. The higher the h, the lower the number of clusters. For example, for $h = 2.5$, we obtain four clusters, as shown in Figure 3.4B, where the clusters obtained are shown as colors in the gene names. Each cluster corresponds to one of the branches of the dendrogram at the given height and contains all the "descendants" of that branch.

3.2.7 K-means Clustering

---

i **Centroid-based clustering**

In *centroid-based clustering*, the user specifies the number K of clusters to be obtained. The algorithm produces K *centroids*, one per cluster, each representing the typical expression profile of the genes included in the cluster. The centroids are typically initialized in an arbitrary way, then an iterative optimization procedure allows them to "adapt" to the genes to be clustered.

We will discuss a specific implementation of centroid-based clustering known as K-means clustering. Compared with UPGMA, this algorithm uses less computing resources, in particular RAM, since it is not necessary to compute and store all pairwise distances between genes, that are needed to build the UPGMA tree. The number of clusters K is predefined by the user. Let us see how it works using again

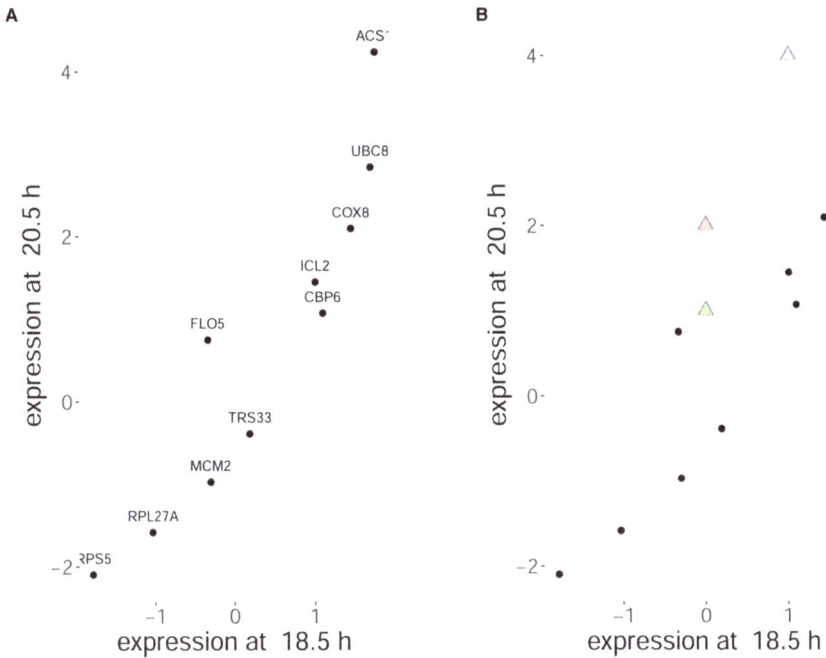

Figure 3.5 (A) Ten genes that we want to cluster are represented as points in a plane (considering only the two latest timepoints of the diauxic shift experiment). (B) $K = 3$ centroids (triangles) have been initiated in randomly chosen positions.

the last two timepoints of the diauxic shift experiment (so as to be able to visualize the genes on a plane) and our ten sample genes.

The algorithm uses K *centroids*: each centroid is the expression profile (thus, in our case, a two-dimensional vector representing expression at $t = 18.5h$ and $t = 20.5h$) of an ideal gene that does not correspond to any real gene but will represent, at the end of the procedure, the average expression profiles of the genes contained in a cluster. Initially the centroids are chosen arbitrarily, for example, with a suitable random extraction. The genes that we want to cluster are shown in Figure 3.5A, and $K = 3$ centroids have been chosen arbitrarily (triangles) in Figure 3.5B.

Each gene is then assigned to the closest centroid (according to our choice of dissimilarity measure, here the Euclidean distance), as shown in Figure 3.6A, and the centroids are moved to the "center of gravity" of the genes assigned to them (i.e., the new centroid coordinates are the average of those of the genes assigned to it – Figure 3.6B).

Since the centroids have moved, the closest centroid to each gene might have changed, so the assignment of genes to centroids is updated (Figure 3.6C), and the centroid position is, in turn, updated to reflect the average of their new genes (Figure 3.6D). At this point, in our simple example, when we try to update the gene assignment, no gene changes centroid. Therefore, the centroid position does not need to be updated, and we say that the process has *converged*. In real cases with thousands of genes and several experimental points, the process will take many iterations to

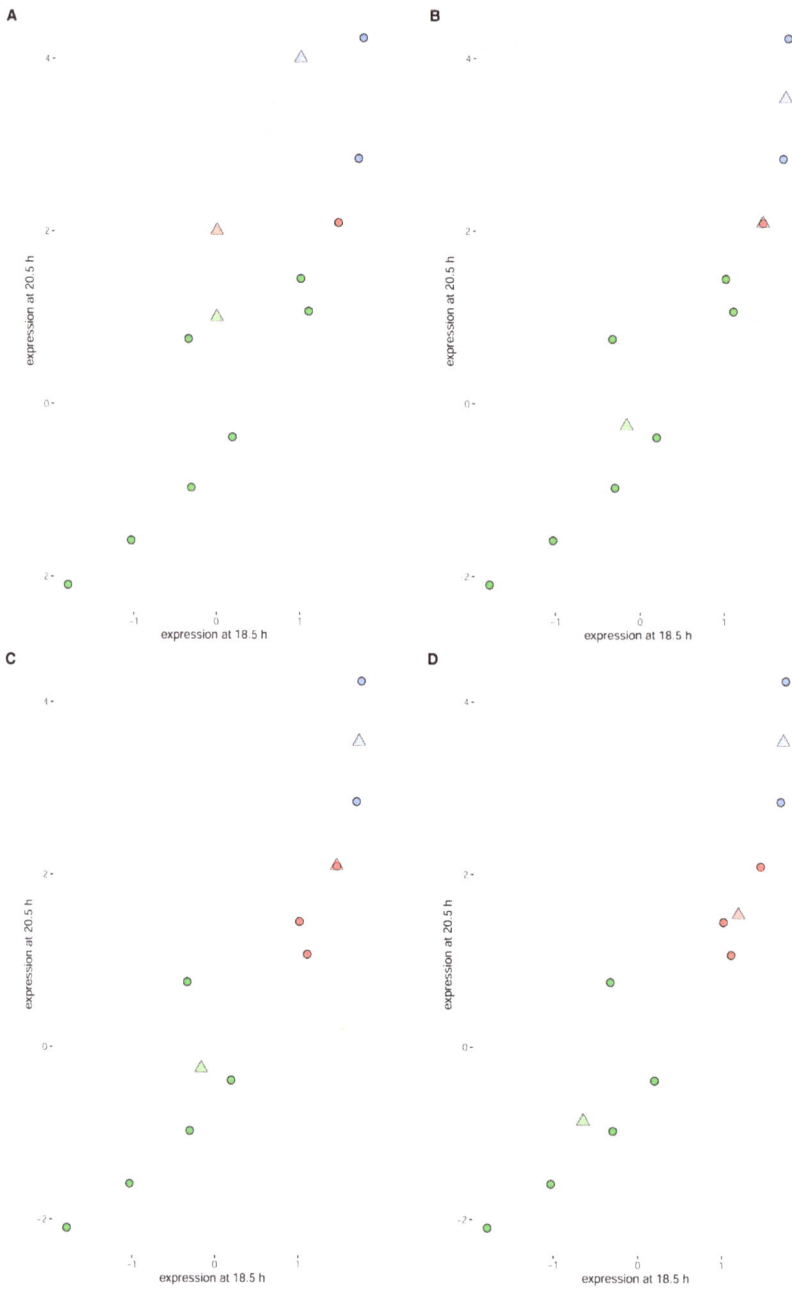

Figure 3.6 (A) Each gene is assigned to the closest centroid. (B) Each centroid is moved to the position corresponding to the average expression of the genes assigned to it. (C) Since the centroids have moved, the genes are reassigned to the closest centroid. (D) The centroids are moved again to coincide with the average expression of the genes assigned to them. At this point, since each gene is already assigned to the closest centroid, the iterative procedure stops (it has reached *convergence*).

converge, but a mathematical theorem guarantees that convergence will eventually be reached. The final assignment of the genes to the centroids is the result of our clustering algorithm. Note that the final result depends on the initial choice of the centroids, and there is no guarantee that different choices will lead to exactly the same clusters. However, in practical cases, the differences usually concern few genes. Procedures have been defined to choose the initial centroids in a deterministic way based on the data, so that the algorithm becomes completely deterministic, just as hierarchical clustering.

3.3   GUILT BY ASSOCIATION AND ENRICHMENT ANALYSIS

3.3.1   Gene Function and Guilt by Association

As briefly discussed in the introduction to this chapter, class discovery is useful, in particular, for functional genomics, that is, the effort to understand the *function* of a gene or a class of genes. For the time being we will define a function of a gene as a biological process that is altered when the activity of the gene is perturbed (this definition will be put in a larger context below when we discuss the Gene Ontology). This is a good definition in that it makes it clear how to experimentally validate the assignment of a function to a gene: We must show that at least in some condition, developmental stage, or cell type, altering the activity of the gene (e.g. by altering it genetically, changing its expression level, or targeting the protein with an inhibitor) leads to a measurable change in the biological process that defines the function. For example, to show that gene A has the function "glucose transmembrane transport" we need to show that by altering the activity of gene A we can alter the transport of glucose across membranes in at least one biological context.

The key concept that allows us to use class discovery to infer gene function, or at least make data-driven, testable hypotheses about it, is called "guilt by association" and can be stated as follows:

> **i   Guilt by association**
>
> Genes showing similar expression patterns across experimental conditions are likely to be involved in the same biological processes

This principle is based on the idea that to perform a certain task (e.g., transporting glucose across a membrane, or building a functional kidney), the organism needs many proteins, which must be turned on simultaneously by gene regulation whenever the task has to be accomplished (i.e., in the appropriate cell type, developmental stage, and environmental conditions). Therefore, if we observe a group of genes which are turned on and off simultaneously, it is reasonable to suspect that this happens because their products are needed together to perform a certain task.

Clustering provides us precisely with groups of genes characterized by similar patterns of expression, so that we can use the guilt by association principle to hypothesize that genes in the same cluster are likely to be involved in similar biological processes. Let us consider, for example, two clusters of genes obtained by K-means clustering

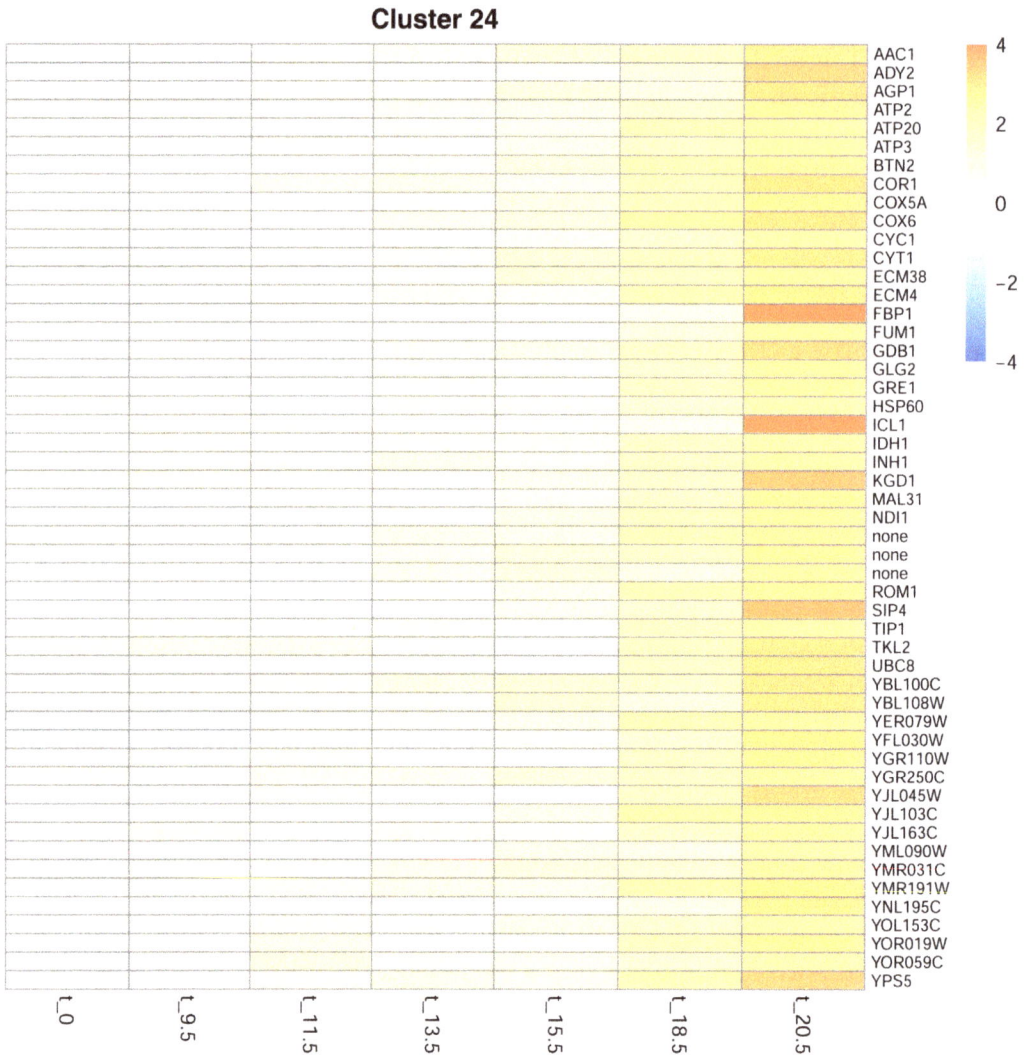

Figure 3.7 K-means clustering of the diauxic shift experiment identifies a cluster of genes strongly upregulated at the later timepoints. The microarray probes marked "none" do not correspond to any known gene.

of the diauxic shift experiment with $K = 60$, so that, on average, each cluster will contain $\sim$100 genes. In Figures 3.7 and 3.8 we show two clusters as heatmaps: The expression level of each gene at each timepoint is expressed through a color scale.

Let us take a closer look at the cluster shown in Figure 3.8, containing genes whose expression decreases at the last two timepoints of the diauxic shift. Simply glancing at the gene names, we see a large proportions of "RPL" and "RPS" genes, that is, genes encoding for ribosomal proteins. We will give a precise meaning to "large proportion" below, but even at first sight we notice that:

1. At least for this cluster, the guilt by association principle seems to hold: These genes, that were grouped together by the algorithm because of their similar

Cluster 36

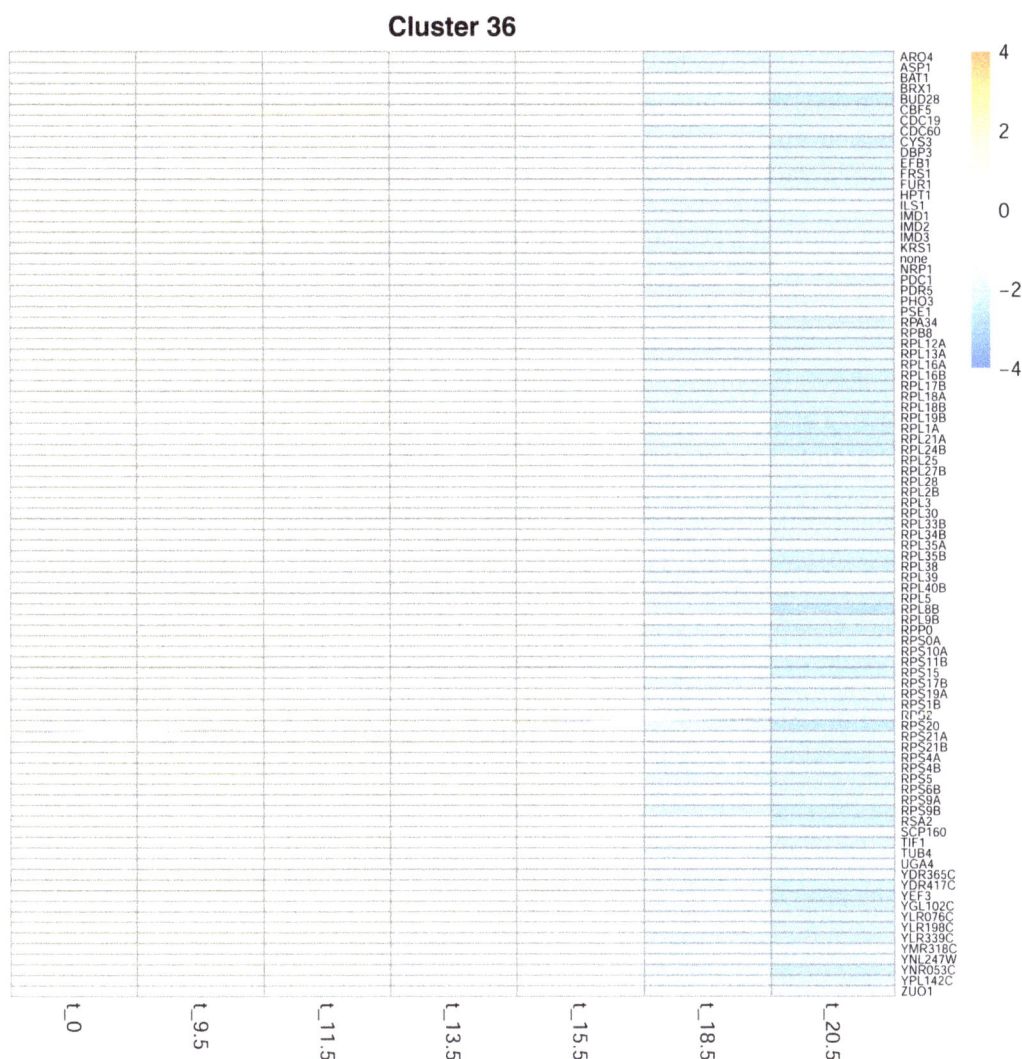

Figure 3.8 Another cluster identified by *K*-means contains genes strongly downregulated at the later timepoints.

expression profiles, do indeed largely participate in the same biological process, that is, the process of translation performed by ribosomes. We learn that toward the end of the diauxic shift, the cells turn off the genes needed to make ribosomes and thus to perform the task of mRNA translation. Many genes are needed to make a ribosome: When ribosomes are less in demand, these genes get turned off simultaneously, and thus they appear in the same cluster in our class discovery analysis.

2. While the genes in this cluster are indeed functionally related, their common function is not the one we would have probably guessed from the experimental design. The diauxic shift represents the metabolic transition from fermentation

to respiration, so we could have guessed that these strongly downregulated genes would be those involved in fermentation. Instead, at least for the cluster we looked at, the common function seems to be a basic cellular function, namely translation. Biologically, this can be interpreted as the yeast cells slowing down their proliferation, and hence needing less translation, as glucose gets depleted.

Therefore, we need a systematic approach to establish *which* common function, if any, characterizes the genes in a cluster. For this we need two main ingredients, that will be introduced in the following subsections:

1. A database containing information about which genes are involved in which function. The Gene Ontology provides such a database with a structure suitable for describing the complexity of the notion of gene function.

2. A quantitative way to establish whether a certain function can be associated to a cluster. Of course it would be unreasonable to require *all* the genes in the cluster to be involved in the same function. We will instead look for functions that are *enriched* in a cluster, that is found among the cluster genes more often that would be expected in a randomly chosen gene list of the same size as the cluster.

To be concrete, and to follow the reasoning that led us here in this chapter, we will describe the procedure used to find enriched functions as applied to a cluster of genes found through class discovery, but the same procedure can be applied to any list of genes derived from a high-throughput experiment for which we have reason to believe they have some functional relatedness. For example, we could analyze the genes that are upregulated in a certain condition, from a class comparison analysis; or the putative regulatory targets of a transcription factor derived from a ChIP-seq experiment (see Chapter 5); or the genes associated to variants found in a GWAS study (see Chapter 6).

3.3.2 The Gene Ontology as a Controlled Vocabulary

If you read somewhere that a gene is involved in "DNA replication," and somewhere else that it is involved in the "replication of DNA," you do not think the two sources are in disagreement, although they used two slightly different phrases. This is because natural language has often many ways of expressing the same concept. As we will need an algorithm running on a computer to process the functions associated to our genes, it is more practical to use a *controlled vocabulary*, that is, to decide once and for all that we will refer to a certain function as "DNA replication" and not as "replication of DNA"[2].

---

[2]It is definitely possible, today, to have a computer understand natural language quite well. Indeed a lot of effort is being put in *text mining*, a process used to derive information from texts written in natural language, for example, article abstracts. Nevertheless, the functional enrichment analysis we will describe is much simpler to implement if we first remove the ambiguities and redundancies of natural language by adopting a controlled vocabulary.

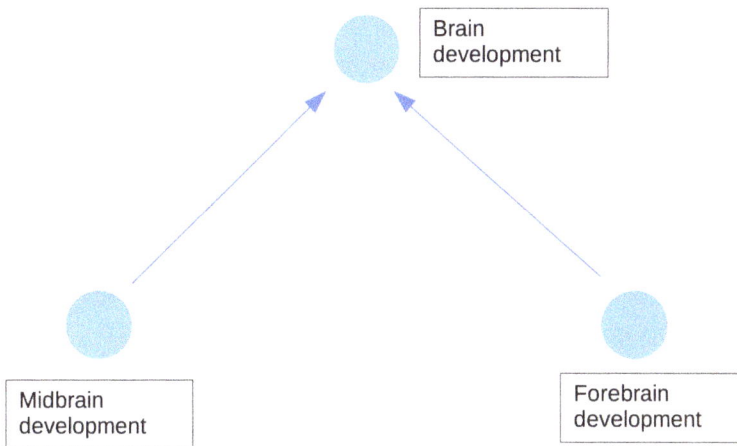

Figure 3.9 The Gene Ontology organizes its terms into a directed graph, in which the edges indicate that a term is part of another term. Thus "Midbrain development" and 'Forebrain development' are parts of "Brain development."

More specifically, we will use a controlled vocabulary embedded within a conceptual structure called an *ontology*. Philosopher Daniel Dennett gave the following definition of an ontology:

> **i Ontology**
>
> The *ontology* of a theory is the catalogue of things and types of things that the theory deems to exist.

The Gene Ontology (GO) represents the ontology of gene function, and is therefore a catalogue of all possible functions (the Gene Ontology *terms*) that a gene can have, and of all the types of function in which these functions are organized. Note that the ontology of a theory evolves in time with the theory: For example, the ether in not anymore in the ontology of physics, while dark matter has been added to it recently.

3.3.3 The DAG Structure

As stated above, an ontology does not list just the things that exist but also the types of things, thus organizing things into classes. For the Gene Ontology, this means for example that "Midbrain development" is a function that will be included in the ontology, but should be considered as part of a more general function "Brain development," which is in turn part of "Nervous system development." The Gene Ontology is thus organized as a graph whose nodes are the gene functions: An edge going from node A to node B indicates that A is a part of B. An example is shown in Figure 3.9.

One would be tempted to assume that the resulting graph would be a tree, with each term branching into several more specific terms. However, a tree structure would be too simplistic, as it would require a term to be part of only one other term. The actual structure of the Gene Ontology is a directed acyclic graph:

> **i** **Directed acyclic graph**
>
> A *directed acyclic graph* (DAG) is a directed graph in which it is not possible to go from one node to itself *following the direction of the arrows*, although it might be possible to do so if the arrow direction is ignored.

When applied to the Gene Ontology, this property allows a term to be a part of more than one other terms, as in the example shown in Figure 3.10.

3.3.4 The Three Domains

The Gene Ontology covers three separate domains, each with its own DAG describing a different type of functional information we might have about a gene:

- *Biological process (BP)*: A biological process represents a specific objective that the organism is genetically programmed to achieve. Examples are brain development, DNA replication, translation.

- *Cellular component (CC)*: The location, relative to cellular compartments and structures, occupied by the gene product when it carries out a molecular function. Examples are nucleus, cell membrane, spindle pole.

- *Molecular function (MF)*: A molecular process that can be carried out by the gene product usually via direct physical interactions with other molecular entities. Examples are catalytic activity, DNA binding, protein transporter activity.

Therefore, each domain tells us something different about how the gene product acts to perform its job in the organism, "Biological process" being the domain that most closely reflects the definition of gene function that we adopted above. For example, consider the human gene *REST*, a transcription factor involved in nervous system development. Its BP annotations include "Negative regulation of neuron differentiation," and the CC ones include, as expected for a transcription factor, "Nucleus", while the MF annotations include "DNA-binding transcription factor activity". So, BP, CC, and MF describe the why, where, and how, respectively, of the biological activity of the gene product.

3.3.5 More on the GO

Each GO term is associated to a GO accession id of the form "GO:" followed by seven digits: For example, the accession GO:0003700 is associated to the term "DNA-binding transcription factor activity" of the MF domain. The GO website contains the structure of the DAG, the precise definition of each term, and the list of genes that have been annotated to the term. This list takes into account the DAG structure, hence a gene annotated e.g. to "Forebrain development" will appear in the lists of

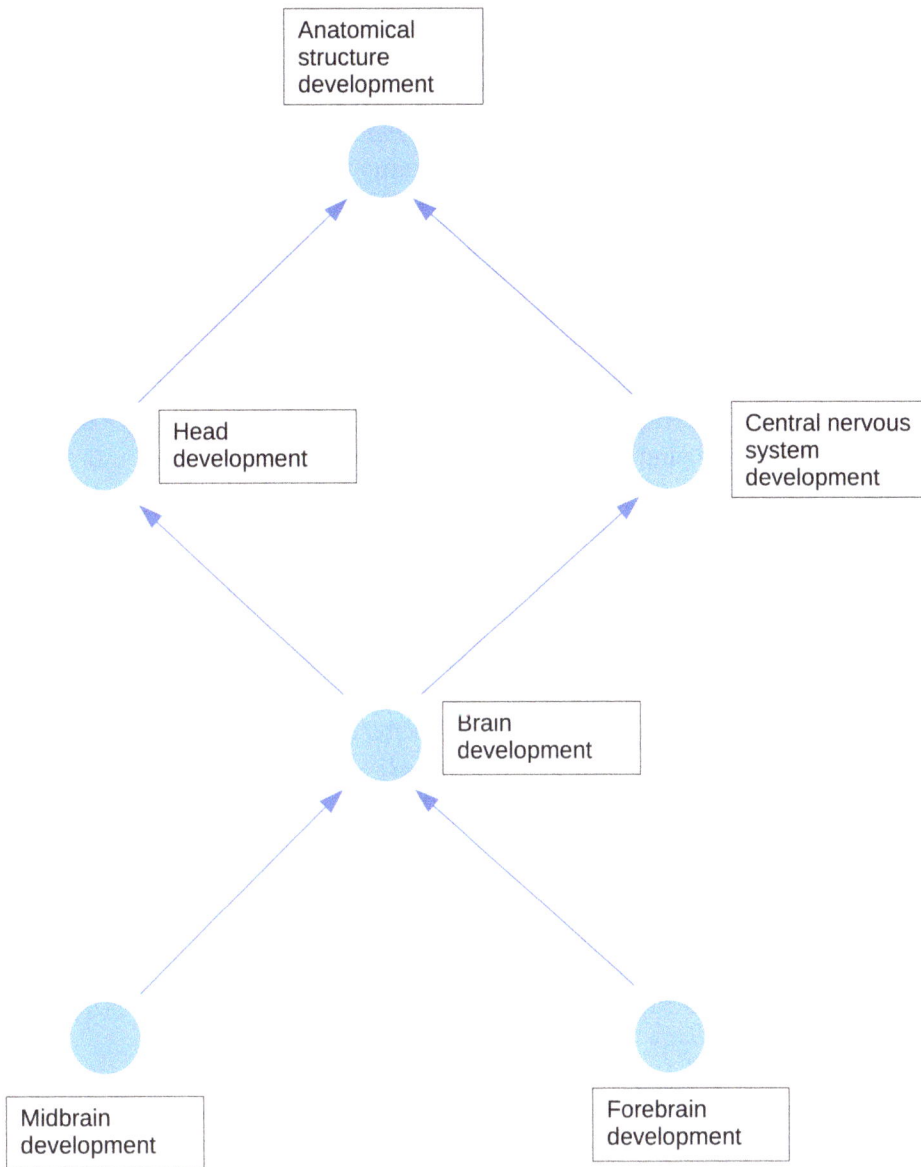

Figure 3.10 The DAG structure of the Gene Ontology allows for a term to be part of several terms, which would not be allowed by a tree structure. Note that, as per the definition of DAG, in this graph it is not possible to go from one node to itself while following the arrows, but it is possible to do so if the arrows are ignored.

genes associated to "Brain development," "Central nervous system development," and so on up to the most general node of the BP DAG, which is the term "Biological process."

Each association between a gene and a GO term is characterized by an *evidence code* which specifies the type of evidence we have that the gene is indeed involved in the function described by the term. Such evidence can have very different levels of confidence, from the "Traceable Author Statement" (TAS) code, implying that the gene/term association can be traced to a specific publication, to codes such as "Inferred from Expression Profile" (IEP), indicating that the annotation is inferred from the timing or location of expression of a gene.

Importantly, the GO is *cross-species*: the same term can be applied to all species for which it is relevant. Thus, a term such as "DNA replication" can have genes from virtually all species annotated to it while a term such as "Brain development" will apply only to the species endowed with a brain. This makes it easier to compare functional analyses performed on genes of different species (e.g., experiments carried out in mouse and human tissues) and to transfer the functional annotation of a gene to the orthologous gene in another species. For example, the chimpanzee *REST* gene is annotated "Negative regulation of neuron differentiation," like the human one, but such annotation is simply derived from that of the human gene, and indeed the evidence code is "Inferred from Biological aspect of Ancestor" (IBA), indicating that the annotation was transferred among related sequences based on common ancestry.

3.3.6 Enrichment Analysis

The Gene Ontology allows us to associate to each gene in a gene list (for example one of our clusters) its known functions, and thus solves the first of the two problems listed at the end of section 3.3.1. The second problem, that is, to determine whether a certain function is *enriched* (or *over-represented*) among the genes in a list, is solved by *enrichment analysis*, a statistical technique based on the exact Fisher test.

---

i **Enrichment**

A functional annotation, such as a GO term, is *enriched* in a list of genes if the prevalence of annotated genes in the list is significantly greater than that expected in a randomly generated list of genes of the same size.

Consider a list of genes, for example, the genes that were classified into the cluster of downregulated genes shown in Figure 3.8. It turns out that 75 among the 89 genes in our cluster have at least a GO BP annotation, and among these 36 (48%) are annotated to the GO term "ribosome biogenesis". Compare such prevalence to what happens in the whole yeast genome, in which 5,276 have at least a GO BP annotation, and 473 (8.97%) of these are annotated "ribosome biogenesis". In a *random* list of yeast genes, we would expect the same fraction, and thus 6.72 genes annotated ribosome biogenesis. Thus, the 36 that we find represent a 36 / 6.72 = 5.35-fold *enrichment*, which would lead us to conclude that this cluster is indeed functionally characterized by involvement in ribosome biogenesis.

As usual, however, we have to make sure that such enrichment did not happen by chance, by computing the probability that a randomly chosen list of 75 yeast genes with BP annotation would contain 36 or more genes annotated "ribosome biogenesis". This is done with Fisher's exact test, which is based on the hypergeometric distribution and is often presented in terms of balls in a box.

Suppose you have a box containing N balls, W of which are white, and pick m balls from the box (without putting them back in the box, that is each ball can be picked at most once). The hypergeometric distribution gives the probability that w or more of the balls you pick are white. This can be easily mapped to our problem: the box is the yeast genome (or more precisely the set of all yeast genes with a BP annotation); the white balls are the genes annotated "ribosome biogenesis"; m represents the number of genes in your random list, and w is the number of genes annotated "ribosome biogenesis" in the list. The Fisher test computes this probability as its P-value; thus a small P-value means that it is very unlikely that a random list of yeast genes would show such an enrichment, and therefore such enrichment is a biologically meaningful property of the cluster, which in particular tells us that guilt by association holds in this case. For our specific case, the P-value is $6.74 \cdot 10^{-19}$, and thus we are very confident of the biological meaning of the enrichment.

We have looked at a specific annotation, but usually the analysis is performed in an unbiased way by using the Fisher test on all GO terms and retaining the ones with significant enrichment. Of course this entails a multiple testing issue (we perform a test for each GO term) which can be taken care of with the same methods discussed in the context of class comparison[3]. The top enriched categories for our cluster are shown in the table:

ID	Description	Set	Genome	P	FDR
GO:0002181	cytoplasmic translation	44/75	206/5276	3.936e-44	1.657e-41
GO:0042254	ribosome biogenesis	36/75	473/5276	6.735e-19	1.418e-16
GO:0042255	ribosome assembly	15/75	80/5276	1.464e-13	2.055e-11
GO:0042273	ribosomal large subunit biogenesis	15/75	127/5276	1.544e-10	1.625e-08
GO:0022618	ribonucleoprotein complex assembly	17/75	184/5276	4.041e-10	3.403e-08
GO:0071826	ribonucleoprotein complex subunit organization	17/75	190/5276	6.702e-10	4.703e-08

Here "Set" and "Genome" indicate the prevalence of each term in the cluster and in the whole genome, respectively, and we have used the false discovery rate to adjust for multiple testing. You can immediately notice that the results appear to be highly redundant: Indeed, some of the top enriched categories are very similar to each other, and their enrichment actually depends on the same genes. This problem is related to the hierarchical nature and the very fine structure of the Gene Ontology; some methods to deal with this problem and to extract more or less independent enriched categories have been developed, but they are beyond the scope of these lectures.

---

[3]The multiple testing issue becomes more severe when, as it is usually the case, the analysis is performed for all clusters, so that the number of tests performed is equal to the number of GO terms multiplied by the number of clusters.

It is interesting to return to the cluster of genes that were upregulated toward the end of the diauxic shift (Figure 3.7) and explore its functional enrichment in the same way. The possible functional characterization of this cluster was not as apparent just by looking at the gene names. The top enriched categories are shown below:

ID	Description	Set	Genome	P	FDR
GO:0009060	aerobic respiration	12/32	93/5276	7.365e-14	1.277e-11
GO:0015980	energy derivation by oxidation of organic compounds	14/32	159/5276	8.482e-14	1.277e-11
GO:0006091	generation of precursor metabolites and energy	15/32	217/5276	3.075e-13	3.085e-11
GO:0045333	cellular respiration	12/32	110/5276	5.853e-13	4.404e-11
GO:0006119	oxidative phosphorylation	6/32	41/5276	1.176e-07	7.078e-06
GO:0022900	electron transport chain	6/32	60/5276	1.208e-06	5.655e-05

Thus, also this cluster is functionally characterized, and guilt by association is confirmed. In this particular case, contrary to what happened for the downregulated cluster, the top enriched GO terms are those that one could have guessed from the experimental design, and the top one is indeed "Aerobic respiration."

3.3.7 The Gene "Universe"

So far we have compared the prevalence of a GO term in a cluster to its prevalence in the whole genome to assess its possible enrichment in the cluster. In many cases the whole genome is not a suitable comparison, and we need to carefully choose the "universe" of genes to which our gene list of interest should be compared. The universe should contain all genes that could have *in principle* ended up in the list. For example, if we are using a microarray that measures only some of the genes of an organism, the prevalence of a GO term in a cluster should be compared to that among the genes represented in the microarray, not to that in the whole genome.

For a more instructive example, suppose we performed a class comparison analysis of an experiment conducted on immune cells. To be specific, suppose we are comparing the transcriptomes of activated and quiescent T cells. T cells, like all cell types, do not express all the genes in the genome: As a rough rule, most cell types express approximately half the genes contained in the human genome. If we want to perform an enrichment analysis of the differentially expressed genes (DEGs) in activated vs. quiescent T cells, we have to compare the prevalence of each GO term among the DEGs to its prevalence among the genes *expressed by T cells*, because only such genes could have ended, a priori, in your DEG list.

A toy example will demonstrate that disregarding this fact could lead to wrong conclusions: Assume the human genome contains 20,000 genes, and 2,000 of them (10%) are annotated "immune response". Assume moreover that T cells express 10,000 genes, including 1,500 (15%) annotated "immune response". Finally, assume we have a list of 100 DEGs, including 15 annotated "immune response." If we compared our DEGs to the whole genome, we would observe a 1.5-fold enrichment of the term "immune response" in our DEGs, and such enrichment would have a P-value of $2.29 \; 10^{-11}$. But the prevalence in the DEGs is the same as the prevalence among genes expressed by T cells, so the enrichment we observe is not a feature of our DEGs, but of genes expressed by T cells. If we compare the DEGs to the "universe" of the genes expressed by T cells, "immune response" is not enriched.

3.3.8 Applications to Functional Genomics

The guilt by association principle suggests a way to use gene expression data to gain new knowledge in functional genomics. Consider a cluster with a strong functional enrichment, such as our downregulated cluster of Figure 3.8. As we saw, not all genes in the cluster are annotated to the most enriched terms, for example, "Cytoplasmic translation." The genes in the cluster that are *not* annotated "Cytoplasmic transla-tion" are however good candidates to be new genes involved in such function, as their expression profile matches that of many genes thus annotated. Therefore, if we are interested in finding new genes involved in cytoplasmic translation, this list would be a very good starting point. Of course specific experimental work would be needed to verify whether any of these genes is really involved in cytoplasmic translation, but the probability of success is certainly increased compared with starting from a random list of genes, or from the entire yeast genome. Computational analysis of transcriptomic data thus provides a *prioritization* of experimental work.

3.3.9 Gene Set Enrichment Analysis

As we discussed above, the gene lists that can be analyzed for functional enrichment are not only clusters found by class discovery: Any gene list produced by a high-throughput measurement technique, and expected to contain functionally related genes, can be analyzed with the Fisher test for enrichment of genes annotated to Gene Ontology terms. For example, it is quite common to use enrichment analysis on the lists of differentially expressed genes found by class comparison. As an example, let us analyze the genes found downregulated in old animals in the zebrafish experiment of Chapter 2 in terms of GO BP functional enrichment: At a FDR of 0.1, we have 21 such genes, and some functional enrichments turn out to be indeed significant (with the usual redundancy)[4]:

---

[4]The number of genes in the set as it appears in the table is smaller than the total number of downregulated genes since it includes only genes with at least one BP annotation and included in the appropriate gene universe, here defined as the genes with expression value above a detection threshold in at least one of the samples.

ID	Description	Set	Genome	P	FDR
GO:0051028	mRNA transport	4/18	46/7523	3.517e-06	0.0009765
GO:0050657	nucleic acid transport	4/18	59/7523	9.62e-06	0.0009765
GO:0050658	RNA transport	4/18	59/7523	9.62e-06	0.0009765
GO:0051236	establishment of RNA localization	4/18	59/7523	9.62e-06	0.0009765
GO:0006403	RNA localization	4/18	67/7523	1.601e-05	0.001083
GO:0015931	nucleobase-containing compound transport	4/18	67/7523	1.601e-05	0.001083

However the enrichment of, for example, mRNA transport is driven by just four genes, while 46 in our gene universe are so annotated. It is natural to ask whether other genes annotated as involved in mRNA transport show some evidence of down-regulation, even if not enough to pass the statistical threshold, or, more generally, whether the genes associated to any given GO term show a coherent pattern of fold changes (e.g., by being mostly up- or downregulated) regardless of the statistical significance of the differential expression of each individual gene. This is the approach taken by *gene set enrichment analysis* (GSEA) [36].

> **i Gene set enrichment analysis**
>
> *Gene set enrichment analysis* (GSEA) determines whether the fold changes of the genes in a gene set (e.g., a GO annotation) are systematically biased toward high or low values, irrespective of the statistical significance of the differential expression of the individual genes in the set.

A preliminary version of GSEA was described in [30] and is easier to understand conceptually, although it presents some drawbacks that were corrected in the current version. Therefore, we will present the preliminary version and refer the reader to [36] for the state-of-the-art version.

First, all genes are ranked by decreasing fold change. Let N be the total number of genes measured, and S the number of genes belonging to a gene set (e.g., annotated to a GO term). Then, a *running score* R is computed by going down the list in the following way:

- set $R = 0$
- for each gene encountered in the sorted list

    - if the gene belongs to S, add $\frac{1}{S}$ to R
    - otherwise, subtract $\frac{1}{N-S}$ from R

- keep track of the value of R after each step

Since all measured genes are in the list, when we reach the end R will be again equal to 0 ($= S \cdot \frac{1}{S} - (N-S) \cdot \frac{1}{N-S}$). If (null hypothesis) the genes in the gene set are distributed randomly along the list, then R will show small fluctuations around zero as we go down the list. However, if most of them have large, positive fold changes, and thus appear toward the top of the list, R will first grow to a positive value much larger than zero, and then gradually decrease to zero. If most genes in the set are downregulated, R will reach a large negative value toward the end of the list before returning to zero. A toy example is shown in the table below, where ten genes are sorted by decreasing fold change and two gene sets with three genes each are considered. $S1$ and $S2$ are binary variables representing whether each gene belongs to each set, $r1$ and $r2$ represent the contribution of each gene to the running score, and $R1$ and $R2$ the corresponding running scores:

	logFC	$S1$	$r1$	$R1$	$S2$	$r2$	$R2$
gene 1	0.9047	0	$-1/7$	-0.1429	1	$1/3$	0.3333
gene 2	0.7753	1	$1/3$	0.1905	1	$1/3$	0.6667
gene 3	0.5709	0	$-1/7$	0.04762	0	$-1/7$	0.5238
gene 4	0.5213	0	$-1/7$	-0.09524	1	$1/3$	0.8571
gene 5	0.516	0	$-1/7$	-0.2381	0	$-1/7$	0.7143
gene 6	0.09894	1	$1/3$	0.09524	0	$-1/7$	0.5714
gene 7	-0.3038	0	$-1/7$	-0.04762	0	$-1/7$	0.4286
gene 8	-0.5318	0	$-1/7$	-0.1905	0	$-1/7$	0.2857
gene 9	-0.5871	1	$1/3$	0.1429	0	$-1/7$	0.1429
gene 10	-0.8362	0	$-1/7$	0	0	$-1/7$	0

The running scores $R1$ and $R2$ are plotted in Figure 3.11 as a function of the gene number. The genes in the set S1 are quite uniformly distributed in the ranked list, so that the running score (blue line) oscillates around zero. Instead, the genes in S2 accumulate at the top of the list, and the running score (red) grows quite far from zero before descending. The *enrichment score* of a gene set is defined as the maximum displacement of the running score from the x axis, and is positive (negative) for gene sets that accumulate near the top (bottom) of the list, and are thus mostly upregulated (downregulated).

Of course we need to associate a P-value to the enrichment score: The best way to do this is to use an empirical approach in which the sample classes (in our case "young" and "old") are randomly redistributed among the samples, and fold changes and enrichment scores are computed with the randomized labels. If this procedure is repeated many times (e.g., 1000 times), we obtain an *empirical probability distribution* of the enrichment score. This distribution reflects the null hypothesis of no enrichment, since the fold changes are biologically meaningless after randomization of the class labels assigned to the samples. Therefore, it can be used to assess whether the enrichment score obtained from the real data without randomization is likely to appear by chance: For example, if the true enrichment score is higher (in absolute value) than 990 out of 1,000 enrichment scores obtained from randomized data, the P-value can be estimated to be 0.01.

However, this procedure requires a large number of samples in each class to generate a large number of permutations. In most transcriptomics experiments, the number

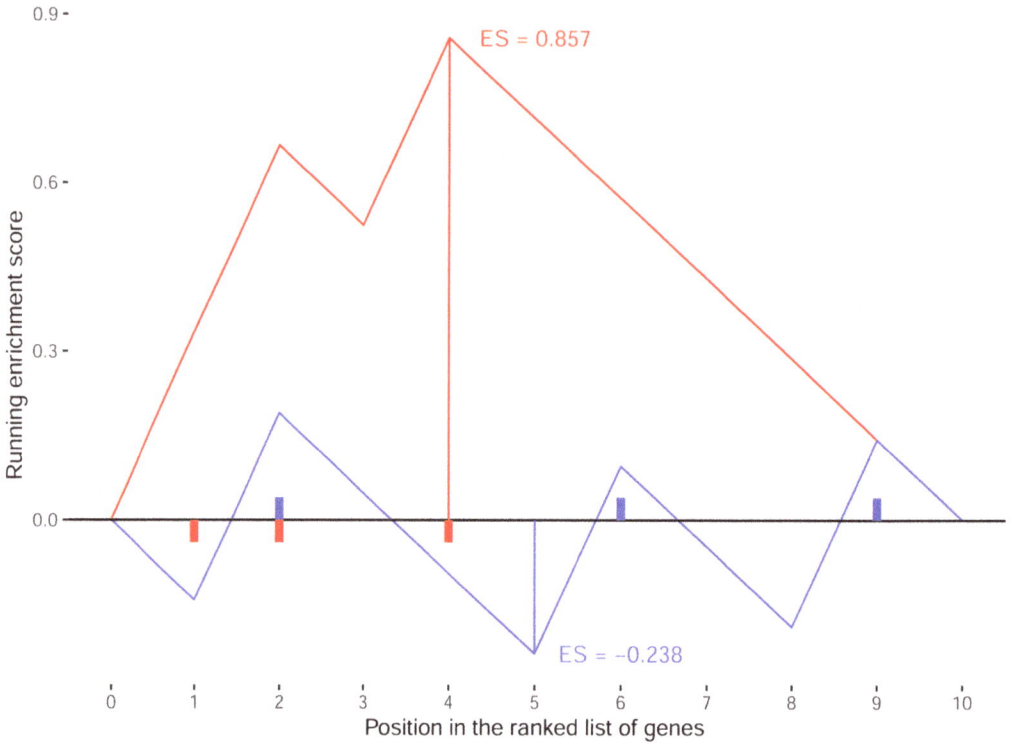

Figure 3.11 Running scores for gene sets $S1$ (blue) and $S2$ (red) in the toy example of gene set enrichment analysis, with the respective enrichment scores. The genes in set $S1$ (blue rectangles) are distributed quite randomly in the list, so that the running scores is never far from zero and the enrichment score is low in absolute value. Instead, the genes in gene set $S2$ (red rectangles) are concentrated at the top of the list, generating a running score which grows to a high value and a high enrichment score.

of replicates for each class is rather small (often as small as 3), and this procedure cannot be applied. In this case, it is possible to compute a different empirical P-value by randomizing the genes instead of the samples, that is, by replacing the gene set under analysis with many random gene sets of the same size. This procedure neglects the fact that the genes in an actual gene set tend to have correlated fold changes, thus the use of random gene sets is not a very satisfactory null hypothesis, and the resulting P values should be considered as indicative. In practice such *pre-ranked* GSEA is the most used, as it is suitable to the sample size of most real transcriptomic experiments. We will use pre-ranked GSEA in our zebrafish experiment.

Let us first look at the distribution of the genes annotated "mRNA transport" (the most significant term in our enrichment analysis) in the list of genes ranked by decreasing fold change, shown in Figure 3.12, where the genes annotated to the GO term are represented as vertical bars. The genes seem to be rather randomly distributed across the fold change values: The enrichment score is -0.285, and the

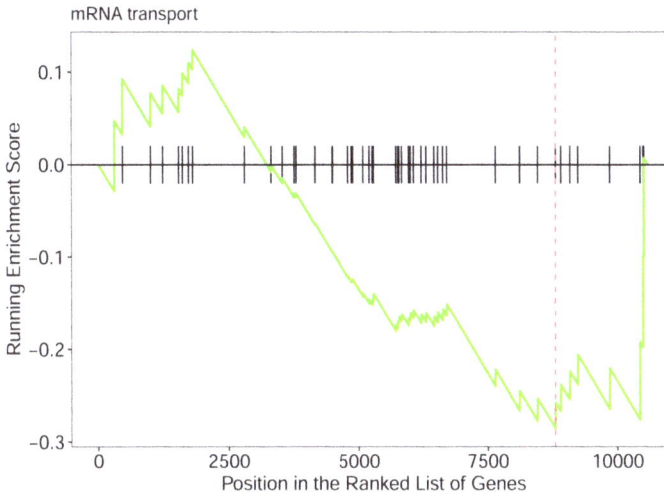

Figure 3.12 Gene set enrichment analysis of the GO annotation "mRNA transport" in the zebrafish brain transcriptomic dataset. The genes annotated "mRNA transport" (vertical bars) do not show a strong bias toward either tail of the distribution, and indeed the enrichment score (dashed red line) is not statistically significant ($P = 0.516$).

P-value as computed by pre-ranked GSEA is 0.516. Thus for this GO term, we have a significant enrichment among downregulated genes due to the 4/18 annotated genes among those significantly downregulated, but no significant global trend towards down- (or up-) regulation of the genes annotated "mRNA transport" as a whole. However, other GO terms show significant trends in the GSEA analysis: For example the fold change of the genes annotated "calcium ion homeostasis" is shown in Figure 3.13, where the enrichment score is -0.557, corresponding to a P-value of $5.14 \cdot 10^{-5}$.

Thus, GSEA can be considered as an alternative way to analyze a gene list from the point of view of gene function. It is complementary to enrichment-based analysis because it does not distinguish between genes that are or are not differentially expressed, but rather looks at the overall behavior of an entire set of functionally related genes.

Pre-ranked GSEA simply requires the values of the fold change for each gene. This implies that it can be extended to any other context in which genes can be ranked, not necessarily by fold change. For example, suppose we are interested in the function of a gene G, and we have transcriptomics measurements across a range of conditions. The guilt by association principle suggests that the genes that are most correlated with G will tend to share G's function, so a possible strategy is:

(1) Rank all genes by decreasing correlation with G.
(2) Use pre-ranked GSEA with functional gene sets (e.g., GO terms), replacing the fold change with the correlation with G.

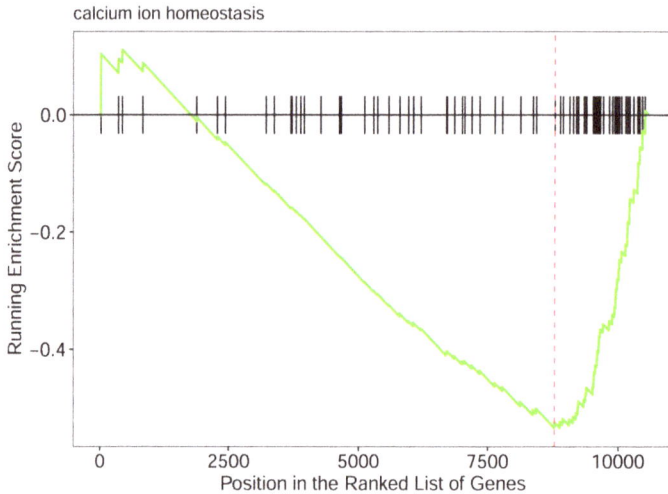

Figure 3.13 Gene set enrichment analysis of the GO annotation "calcium ion homeostasis" in the zebrafish brain transcriptomic dataset. The genes annotated "calcium ion homeostasis" accumulate in the right tail of the rank distribution, and the enrichment score (dashed red line) is statistically significant ($P = 5.14 \ 10^{-5}$).

A positive and significant enrichment of a GO term suggests that the genes annotated to it tend, as a whole, to be strongly correlated with G, and thus that G might share their function. Also negative enrichments could be informative, by identifying the GO terms whose genes tend to be *anticorrelated* with G.

3.4 CLASSIFYING SAMPLES

3.4.1 Clustering the Columns

The same class discovery methods, and in particular clustering algorithms, can be used to answer completely different biological questions when applied to the columns instead of the rows of our expression matrix, that is, to discover classes of experimental samples characterized by similar transcriptomes. Among many applications of this type of analysis, two are most prominent in the literature:

1. The classification of pathological samples (especially tumors) based on the transcriptome can make possible a molecular classification of pathologies that could improve and integrate the classification based on macroscopic features, and thus possibly be useful to understand and/or treat the disease.

2. In single-cell transcriptomics, clustering of the cells is often the first step to understand the composition of the sample in terms of cell types.

Here, we will discuss the first of these applications, while clustering of single-cell expression data will be discussed in Chapter 4.

3.4.2 Molecular Classification of Tumor Samples

The Cancer Genome Atlas Program (TCGA) collected several types of high-throughput data from $\sim$10,000 human tumor samples spanning $\sim$30 cancer types. Among the data collected are transcriptomic assays, mostly performed by RNA-sequencing. Here we will consider 515 samples of low-grade glioma to demonstrate class discovery as applied to samples, rather than genes. Of course, the same dataset could also be used to perform class discovery on the genes, and such analysis might produce some insight on the molecular biology of this type of tumor.

To classify the samples based on the transcriptome, we use correlation-based hierarchical clustering considering the 500 autosomal genes most variable across the samples[5]. To keep the analysis as simple as possible, we cut the tree so as to obtain two clusters, composed of 204 and 311 samples, and shown as a heatmap in Figure 3.14.

Note that most clustering algorithms, and in particular hierarchical clustering used here, will always produce the desired number of clusters. Thus the fact that our tumors could be classified into two clusters based on their transcriptome is not informative *per se*. To determine whether these clusters contain biologically/clinically meaningful information we need to show that they differ in some relevant property that is not gene expression.

An obviously relevant feature of a tumor is its degree of aggressiveness, or the prognosis of the patient. Since these are retrospective data, we do have information on the progression of these tumors, and in particular we know how long each of these patients survived after surgery. Thus, an interesting question is whether the survival time of the patients in the two clusters is significantly different. If we can show that, we have shown that classifying tumors based on their transcriptome can reveal groups of tumors of different prognosis.

3.4.3 Survival Analysis

To determine whether our two clusters of patients differ in tumor aggressiveness, as measured by the survival time of the patients, we need a set of specific statistical techniques collectively known as *survival analysis*.

> **i Survival analysis**
>
> *Survival analysis* is a group of statistical methods used to analyze the probability that an event will occur as a function of the time elapsed (the *survival time*) from a given start of observation.

For example, in oncology, the event could be the death of the patient (in which case "survival" is meant literally, and is often referred to as "overall survival"), or a recurrence of the tumor (in this case we talk about "recurrence-free survival"); and the start of observation could be surgery or diagnosis. As we have seen for other methods, also these apply to problems in disparate contexts: For example, in manufacturing,

---

[5]We remove genes residing on the X and especially the Y chromosome because they might have high variability simply because samples of both sexes are included.

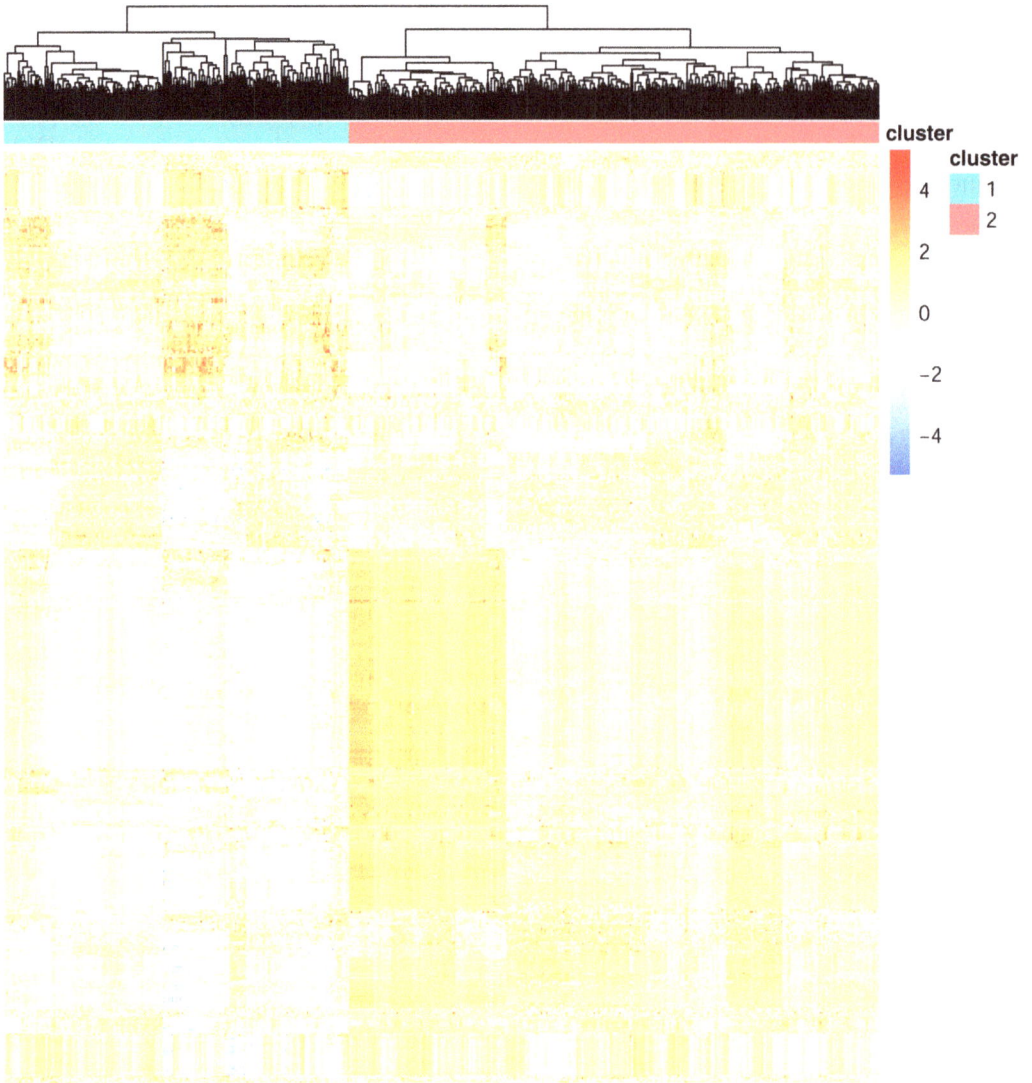

Figure 3.14 Hierarchical clustering of 515 samples of low-grade glioma based on the expression levels of the 500 most variable autosomal genes.

the event of interest might be the failure of a piece of equipment and the start of observation its production or sale. To be concrete, in the following we will refer to cancer, the event being "death" and surgery considered as the start of observation.

3.4.4 Censoring

The methods of survival analysis are devised in such a way as to take into account also cases for which the patient is observed for a certain amount of time without the event being observed. This is important because such patients are often present in real-life datasets and informative: Indeed, if we observe a patient for two years and do not observe the event, we do not know their exact survival time, but we know that it is at least two years.

> **i Censoring**
>
> In *survival analysis* a subject that was observed for a certain amount of time without observing the event of interest is called *censored*. The methods of survival analysis are specifically devised to make use of the information derived from censored subjects.

A survival dataset with censoring specifies two variables for each patient, *time* and *event*. The latter is a binary variable specifying whether the event was observed or not, while the meaning of the *time* variable depends on the value of *event*: If *event* is 1, *time* is the time elapsed from surgery to death; if *event* is 0, it is the time after which observation was terminated with the patient being alive. Such termination can occur for many different reasons (a patient moving to a different region, or the expiration of the time allotted for the study, etc.). For example, these are the overall survival data for some patients included in the low grade glioma sample analyzed above:

Case	Time (days)	Event
TCGA-CS-4941	234	1
TCGA-DB-5274	2289	0
TCGA-DB-A64W	438	1
TCGA-E1-5303	2052	1
TCGA-FG-6691	1257	0
TCGA-S9-A7IX	819	1
TCGA-S9-A7J3	629	0

It is quite obvious from these data that although we did not observe the event for the second case, we can certainly tell that this was a much less aggressive tumor than the first one: This is why it is necessary to devise methods able to take into account both observed and censored cases when analyzing survival data.

3.4.5 Kaplan-Meier Survival Curves

The most well-known tool in survival analysis is the *Kaplan-Meier curve*, which estimates the *cumulative survival probability*.

> **i Cumulative survival probability and Kaplan-Meier method**
>
> The *cumulative survival probability* $S(t)$ is the probability that an individual will be alive after a time t from the start of the observation. It can be estimated from a survival dataset with censoring using the *Kaplan-Meier method*.

Note that if we had observed the event and recorded the corresponding time for all cases, estimating $S(t)$ would be trivial, as it would simply be the fraction of subjects for whom the variable *time* is greater then t. It is the need to take into account information from censored subjects that forces us to use a more complex estimation procedure, the Kaplan-Meier method, which we will illustrate for the seven cases shown above.

- First, we order the table by increasing time:

Case	Time (days)	Event
TCGA-CS-4941	234	1
TCGA-DB-A64W	438	1
TCGA-S9-A7J3	629	0
TCGA-S9-A7IX	819	1
TCGA-FG-6691	1257	0
TCGA-E1-5303	2052	1
TCGA-DB-5274	2289	0

- Then, we consider the intervals between times: 0-234, 235-438, 439-629, 630-819, 820-1257, 1258-2052, 2053-2289.

- For each of these intervals we compute:

  - The number d of events: This is 0 if the end of the interval is a censorship, one if it is an event, >1 if several events happen at the same time.
  - The number n of individuals that were observed (i.e., neither dead nor censored) during the same interval.

- The ratio d/n is our estimate of the probability of the event during the interval.

- Hence $1 - d/n$ is our estimate of the probability of surviving during the interval.

In our case we have the following results

Interval	n_event	n_obs	event_probability	survival_probability
0-234	1	7	0.1429	0.8571
235-438	1	6	0.1667	0.8333
439-629	0	5	0	1
630-819	1	4	0.25	0.75
820-1257	0	3	0	1
1258-2052	1	2	0.5	0.5
2053-2289	0	1	0	1

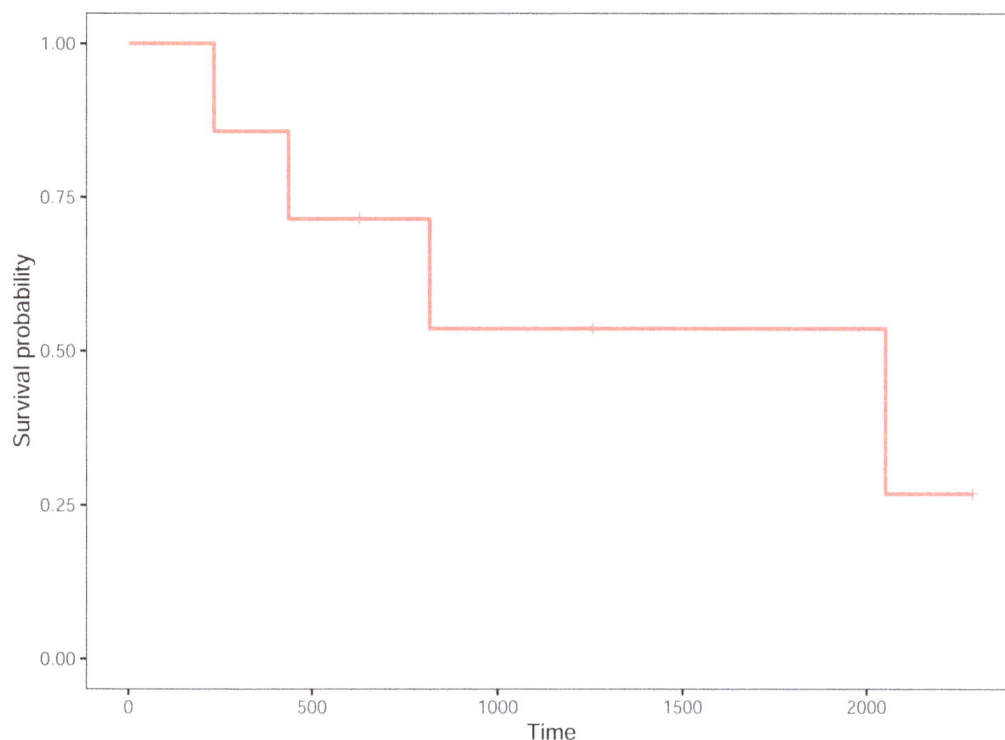

Figure 3.15 Cumulative survival probability estimated with the Kaplan-Meier method for seven low-grade glioma cases.

Note that the number of subjects observed decreases steadily with time due to both deaths and censorships. Finally, the *cumulative survival probability* at the end of each interval is the product of the survival probability for the interval and of all previous intervals. Indeed, to survive until, for example, time 629, a patient has to survive during the intervals 0-234, 235-438, and 439-629, so that the cumulative probability is the product of the survival probabilities during these three intervals:

Interval	n_event	n_obs	event_probability	survival_probability	Cumulative
0-234	1	7	0.1429	0.8571	0.8571
235-438	1	6	0.1667	0.8333	0.7143
439-629	0	5	0	1	0.7143
630-819	1	4	0.25	0.75	0.5357
820-1257	0	3	0	1	0.5357
1258-2052	1	2	0.5	0.5	0.2679
2053-2289	0	1	0	1	0.2679

The cumulative survival probability thus estimated is shown in the Kaplan-Meier curve (Figure 3.15), in which censoring times are shown by small vertical tickmarks. This is our estimate of the survival probability of a patient with low-grade glioma

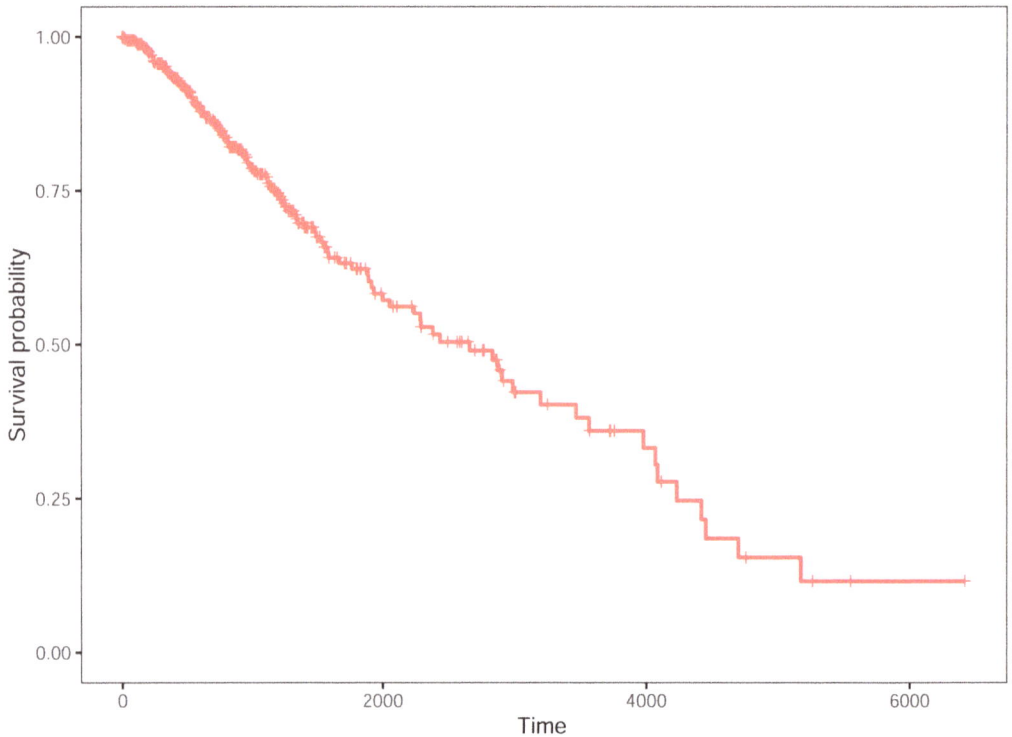

Figure 3.16 Cumulative survival probability estimated with the Kaplan-Meier method for 515 low-grade glioma cases.

based on a very small sample of seven subjects. If we use all the 515 subjects in the dataset we get a much better estimate, shown in Figure 3.16.

This is potentially useful information, but our original goal was to determine whether the two clusters generated from the tumor transcriptome corresponded to patients with *different* survival probabilities, indicating different tumor aggressiveness, or response to therapy. Therefore, we need to generate separate Kaplan-Meier curves for the two clusters and see whether they differ. We obtain the two curves shown in Figure 3.17.

This comparison suggests that patients in cluster 2 have better prognosis, that is, longer survival, than those in cluster 1. For example, 2,000 days after surgery, a patient in cluster 1 has a probability of being alive of 0.459 while the same probability is 0.651 for patients in cluster 2. As usual, we need to make sure this difference in survival is not due to chance, that is, we have to formulate the null hypothesis "The cumulative survival probability for the two clusters is the same" and devise a test to compute a P-value, that is, compute the probability of observing the difference seen between the two curves if the null hypothesis is true. Several methods are available to test this null hypothesis, which we will not describe in detail. Using one of them, the log-rank test, on our two survival curves, we obtain a P-value of $6.07 \ 10^{-7}$. Therefore, we can be quite confident that the difference we observe is not due to chance, and

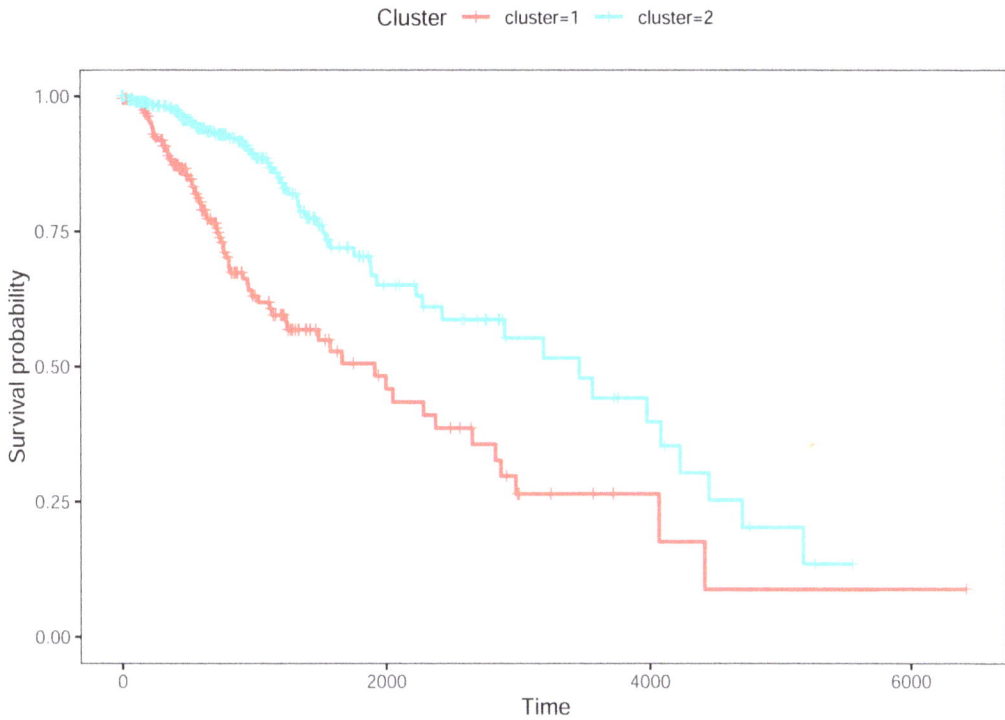

Figure 3.17 Cumulative survival probabilities of the two transcriptome-based clusters of low-grade glioma patients.

thus that transcriptomics can indeed classify patients of low-grade glioma into groups with significantly different prognosis.

FURTHER READING

Ashburner, M., Ball, C., Blake, J., Botstein, D., Butler, H., Cherry, J., Davis, A., Dolinski, K., Dwight, S., Eppig, J., Harris, M., Hill, D., Issel-Tarver, L., Kasarskis, A., Lewis, S., Matese, J., Richardson, J., Ringwald, M., Rubin, G. & Sherlock, G. Gene ontology: Tool for the unification of biology. The Gene Ontology Consortium. *Nat Genet.* **25**, 25–29 (2000).

Eisen, M., Spellman, P., Brown, P. & Botstein, D. Cluster analysis and display of genome-wide expression patterns. *Proc Natl Acad Sci USA.* **95**, 14863–14868 (1998).

Mootha, V., Lindgren, C., Eriksson, K., Subramanian, A., Sihag, S., Lehar, J., Puigserver, P., Carlsson, E., Ridderstråle, M., Laurila, E., Houstis, N., Daly, M., Patterson, N., Mesirov, J., Golub, T., Tamayo, P., Spiegelman, B., Lander, E., Hirschhorn, J., Altshuler, D. & Groop, L. PGC-1alpha-responsive genes involved in oxidative phosphorylation are coordinately downregulated in human diabetes. *Nat Genet.* **34**, 267–273 (2003).

Subramanian, A., Tamayo, P., Mootha, V., Mukherjee, S., Ebert, B., Gillette, M., Paulovich, A., Pomeroy, S., Golub, T., Lander, E. & Mesirov, J. Gene set enrichment analysis: A knowledge-based approach for interpreting genome-wide expression profiles. Proc Natl Acad Sci USA. **102**, 15545–15550 (2005).

Tavazoie, S., Hughes, J., Campbell, M., Cho, R. & Church, G. Systematic determination of genetic network architecture. *Nat Genet.* **22**, 281-285 (1999), http://dx.doi.org/10.1038/10343.

Yersal, O. & Barutca, S. Biological subtypes of breast cancer: Prognostic and therapeutic implications. *World J Clin Oncol.* **5**, 412–424 (2014).

CHAPTER 4

Transcriptomics, Part III: Single-Cell RNA-Sequencing

4.1 INTRODUCTION

In the last decade, the relevance of transcriptomic studies was hugely increased by technological progress that allowed us to measure the transcriptome of an individual cell. This is especially significant since most of the biological samples we are interested in (such as normal or diseased human tissues, or tumors) are composed of cells of several different types. Gene expression as measured by classic ("bulk") transcriptomic assays is thus an average over all the cell types present in the sample, weighted by their relative abundance. Therefore, when we compare transcriptomic samples, for example, by class comparison, the differences we see are the combined result of differences in gene regulation between the two samples, and differences in cell type composition. Disentangling these two effects in bulk transcriptomic assays is a difficult problem, which is partially solved by *deconvolution methods*: Single-cell transcriptomics provides a direct solution.

More generally, single-cell transcriptomics allows us to determine all the cell types present in a given biological samples, and to study the transformation of one cell type into a different one, for example, in the process of differentiation. Thus, single-cell transcriptomics has become a crucial tool in the study of development. Understanding cell-type composition, and its dynamics, is also of fundamental importance in cancer biology: First, to understand the complex interactions between cancer cells and their microenvironment, composed of a variety of non-cancerous cells; and second, because single-cell transcriptomics allows us to study how tumor evolve, in particular, to gain resistance to therapy.

The most basic task in the analysis of single-cell transcriptomic assays is the identification of all the cell types present in a sample. This is typically done by a combination of *dimensional reduction* and clustering algorithms. Dimensional reduction will be introduced in this chapter. The clustering algorithms introduced in Chapter 3 can be and are used also for single-cell data, but specific, *graph-based clustering*

DOI: 10.1201/9781003449928-4

algorithms are used more often and will also be introduced in this chapter. The biological interpretation of the clusters relies on *marker analysis*, which is based on class comparison applied to cell types. Finally, we will briefly introduce *trajectory inference* methods, used to reconstruct how the dynamics of gene regulation guides the transformation of one cell type into a different one, for example, in the process of differentiation, or in cancer evolution.

4.2 DIMENSIONAL REDUCTION

Many types of data in modern biology have *high dimensionality*, meaning that a data point is described by a large number of values, as previously discussed in the context of transcriptomics in Chapter 3: For example, a sample in a human transcriptomic assay in which we measure the expression of all protein-coding genes can be thought of as a point in a ~20,000-dimensional space. This high-dimensionality creates a series of problems in the analysis of these data, the most obvious one being how to represent the data in a way that can be understood and interpreted by our brain.

> **i Dimensional reduction**
>
> Dimensional reduction allows representing high-dimensional data in a space of lower dimensionality, while preserving as much as possible the meaningful properties of the original data, and is used to allow data visualization and to simplify data analysis.

We will first introduce a linear method for dimensional reduction, principal component analysis (PCA), and then briefly discuss newer, non-linear methods (tSNE and UMAP) that are often used in the analysis of single-cell transcriptomic data.

4.2.1 Principal Component Analysis

The classical method for performing dimensional reduction is *principal component analysis* (PCA):

> **i Principal component analysis**
>
> Principal component analysis (PCA) performs a change of coordinate system in the original D-dimensional data, from the original coordinates to new coordinates (the *principal components*) that are *linear combinations* of the original ones, and such that:
>
> (1) The first principal component accounts for the maximum amount of variance that can be accounted for using a single coordinate.
> (2) The second principal component is orthogonal to the first one and accounts for the maximum amount of residual variance (i.e., not accounted for by the first PC) that can be accounted for using a single coordinate orthogonal to the first PC.
> (3) And so on until the D-th PC.

A

B

C

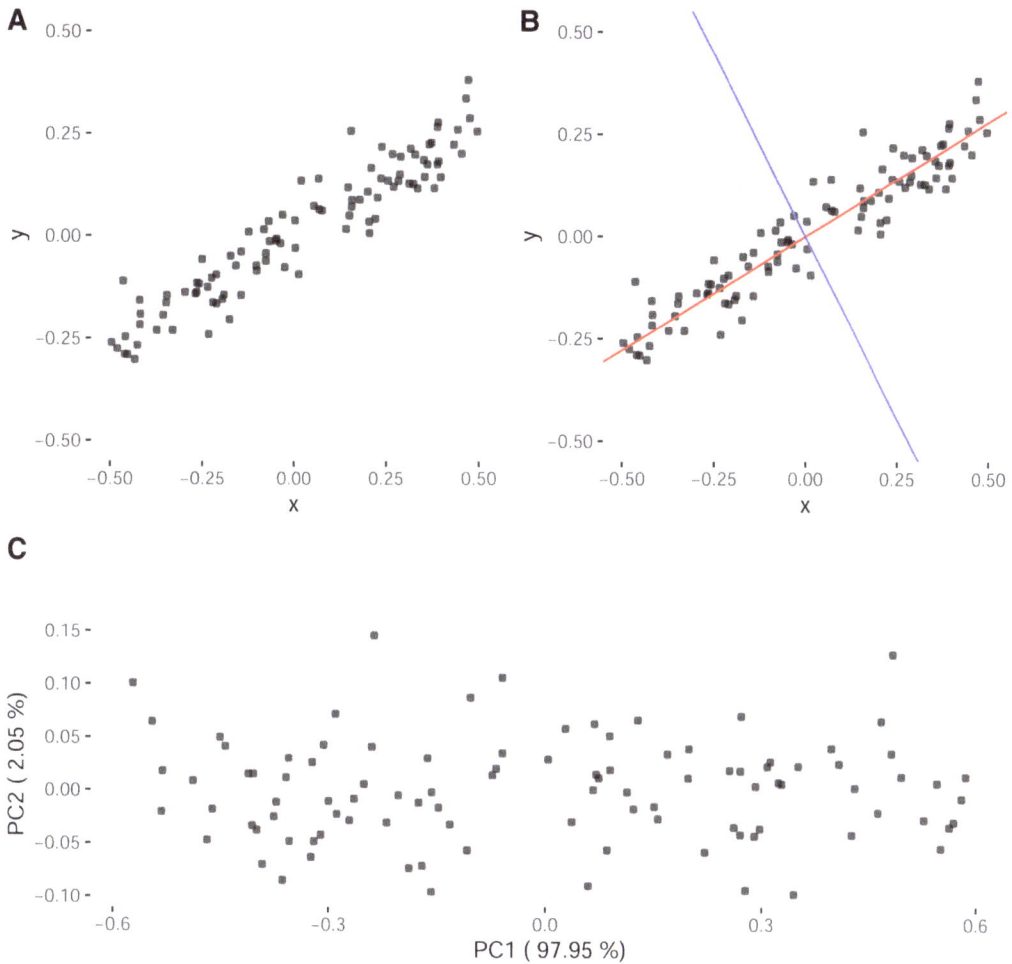

Figure 4.1 (A) A 2-dimensional dataset with 100 data points. The points tend to lie approximately on a straight line of slope ∼0.5. (B) The new axes found by principal component analysis (first principal component in red, second in blue). (C) The data shown in the new axes, with the percentage of variance explained by each component. Since here we are not actually performing dimensional reduction, as the original data were two-dimensional, there are only 2 PCs, and together they explain 100% of the variance.

Let's consider an example with $D = 2$ and $N = 100$ data points, shown in Figure 4.1A. Looking at the data, we see that while they are indeed two-dimensional, they tend to lie approximately on a straight line of slope ∼0.5. So, if we used a coordinate system in which one axis had slope ∼0.5 and the other axis was orthogonal to it, most of the variance would be accounted for by the first coordinate. This is precisely what PCA does. Figure 4.1B shows the new axes in red and blue (first and second PC). At this point, we can show the data points in the new coordinate system

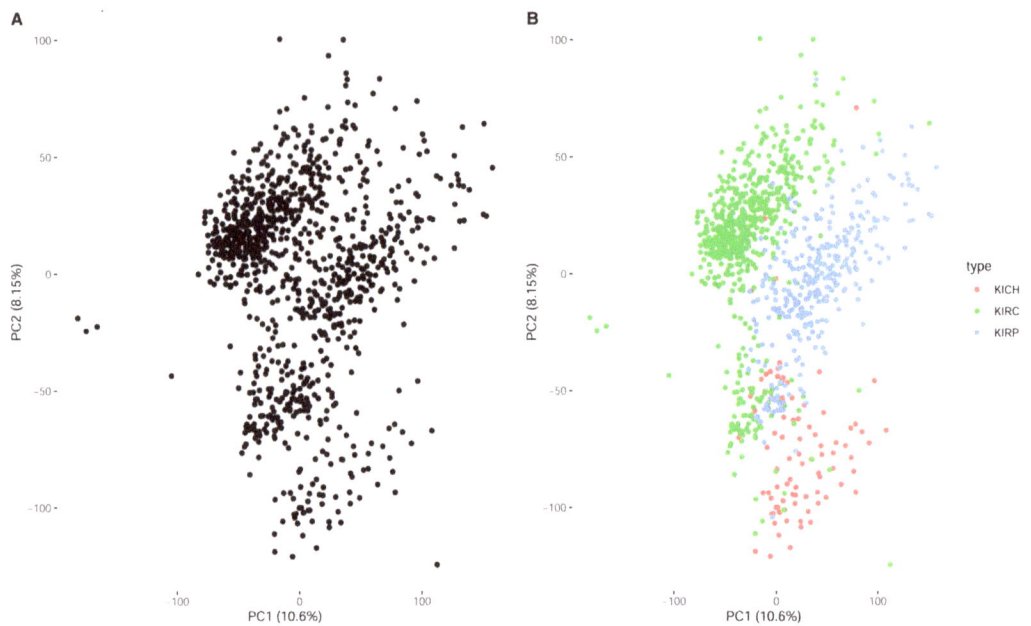

Figure 4.2 (A) Dimensional reduction by PCA of the transcriptomes of 1,020 kidney cancer patients from the TCGA project. (B) The samples have been colored by subtype of kidney cancer: The fact that colors are fairly well separated demonstrates that the first two PCs indeed capture much of the biologically relevant variance.

(Figure 4.1C), where within parentheses the fraction of variance explained by the two PCs is specified.

Note that the total number of PCs is equal to the original dimensionality, $D = 2$, of the data, and together the two PCs explain 100% of the variance. If we use a number of PCs equal to the original dimensionality of the data, we are simply changing the coordinate system, thus explaining 100% of the original variance without actually achieving any dimensional reduction. However, when the original D is greater than 2, we can use the first two PCs to display our data on a plane, knowing that the coordinates we use capture as much variance as it is possible to capture using only 2 coordinates. In this sense PCA is the best possible dimensional reduction that can be achieved through a linear change of coordinates.

4.2.2 An Example

As an example of the application of PCA to transcriptomics, consider the gene expression data for kidney cancer obtained by the TCGA project. These are (bulk) RNA-seq data of 1,020 tumor samples, in which the expression of 20,531 genes was measured. Thus each sample is described by 20,531 expression values, that is, by a point in a 20,531-dimensional space. We can use PCA to represent the samples in a two-dimensional space (Figure 4.2A): The first two PCs explain, respectively, 10.6% and 8.15% of the variance.

Thus the 1,020 samples have been represented in a plane. We can determine if such representation carries biologically useful information by noting that the samples are actually classified by TCGA into three subtypes of kidney cancer: kidney chromophobe (KICH), kidney renal clear cell carcinoma (KIRC), and kidney renal papillary cell carcinoma (KIRP). Coloring the dots according to this classification we see that tumors of the same type tend indeed to be close in the two-dimensional PC space. Therefore the first two PCs indeed capture biologically relevant variance.

4.2.3 Non-linear Dimensional Reduction

PCA is a linear method (in that the PCs are linear combinations of the original coordinates). Sometimes, however, most of the variance of the data is concentrated along a few dimensions, but in a non-linear way. Look, for example, at the data points shown in Figure 4.3A: Most of the variance is indeed concentrated along a one-dimensional line, which however is not a straight line. Therefore, as shown in Figure 4.3B, PCA cannot capture the fact that the data lie on such a line, and the data in the PC axes look very much like the original ones (Figure 4.3C). There is indeed one "dimension" that explains most of the variance, except it is a curve that cannot be expressed as a linear combination of x and y.

Non-linear methods of dimensional reduction have been developed to deal with these cases (isomaps, locally linear embedding, Laplacian eigenmaps, and many others). In single-cell transcriptomics, two methods are very popular, namely *t-distributed stochastic neighbor embedding* (tSNE) and *uniform manifold approximation and projection* (UMAP). A thorough mathematical explanation of these methods is outside of the scope of these lectures, and we will limit ourselves to a description of the basic principles behind them.

tSNE was introduced in [40], and like many of the methods described in these lectures, did not originate in a biological context, but was presented as a general method to visualize high-dimensional data, with examples taken from image analysis. Given two points x_i and x_j in the original space, a similarity p_{ij} is defined based on their Euclidean distance. Importantly, p_{ij} decays exponentially with the square of the Euclidean distance between i and j, so that p_{ij} is significantly greater than zero only for pairs that are very close in the original space. Each point x_i in the original space is then mapped into a point y_i in the dimensionally reduced space, and a similarity q_{ij} is defined in this space. The points y_i are chosen so as to maximize the concordance between p_{ij} and q_{ij}[1].

Two main differences between tSNE and PCA (besides the non-linear character of tSNE) should be noted:

(1) In tSNE, the dimensions of the dimensionally reduced space are equivalent to each other, as opposed to PCA in which the PCs are numbered in order of decreasing variance explained.

---

[1]More precisely, this description applies also to tSNE's predecessor, called SNE. While in SNE also q_{ij}, like p_{ij}, decays exponentially with the distance between y_i and y_j, in tSNE q_{ij} depends on the distance through a Student's t distribution, a modification that turns out to produce better low-dimensional visualization (see the original paper [40]).

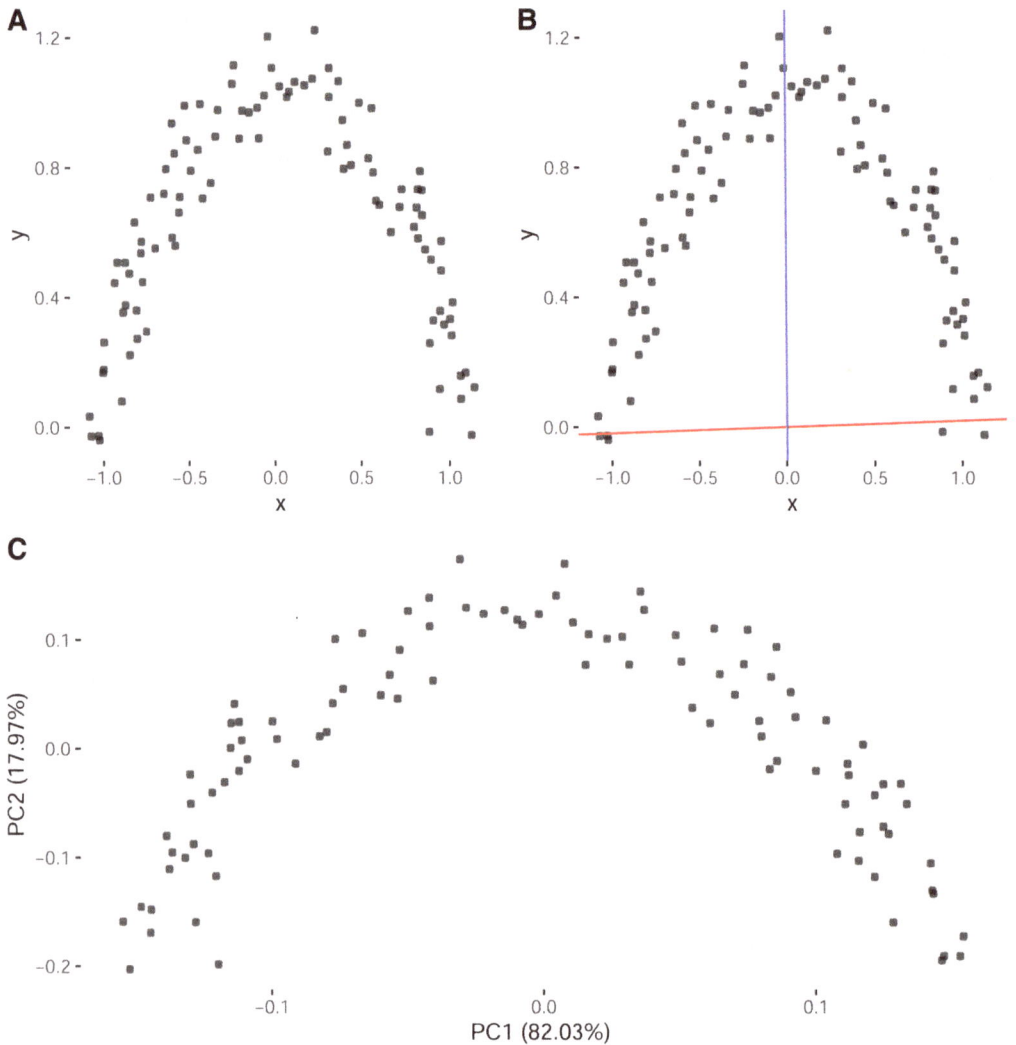

Figure 4.3 (A) A two-dimensional dataset with 100 data points. The points tend to lie approximately on a line, which however is not a straight line. (B) The new axes found by principal component analysis (first principal component in red, second in blue). (C) The data shown in the new axes look very much like the original ones: As a linear method, PCA cannot capture the fact that the data lie on a curve.

(2) While PCA is better at representing the large-scale features of the data (i.e., at placing highly dissimilar points far apart in the dimensionally reduced space), tSNE is better at smaller scales, that is, at placing highly similar points close together in the dimensionally reduced space.

When applying tSNE on the same kidney cancer data that we used to illustrate PCA, we obtain the two-dimensional representation shown in Figure 4.4. As advertised, tSNE separates the three types of kidney cancer more clearly than PCA (and

Figure 4.4 Two-dimensional representation of the kidney cancer samples from the TCGA using tSNE achieves excellent separation of the three known subtypes, although some samples from all three groups get grouped together in what appears to be a fourth subtype.

suggests the existence of a separate group of tumors including samples classified in all the three groups, a potentially interesting lead).

UMAP [29] is another method for non-linear dimensional reduction, very popular in the field of single-cell transcriptomics, although it is again a very general method applicable to any high-dimensional dataset. It is based on Riemannian geometry and algebraic topology, and thus the theory behind it is even more outside of the scope of these lectures than that of tSNE. We will just mention that the two properties that distinguish tSNE from PCA, listed above, equally apply to UMAP.

4.3 THE ANALYSIS OF SINGLE-CELL GENE EXPRESSION DATA

4.3.1 Cell Clusters and Cell Types

We will now describe the main steps of the typical analysis of single-cell transcriptomic data, which heavily rely on the dimensional reduction techniques described above. The main goals of the analysis are:

(1) To identify clusters of cells sharing similar transcriptomes and visualize them in reduced dimension. These clusters are interpreted as the cell types present in the sample.

(2) To map the clusters/cell types into biologically known cell types by analyzing *markers*, that it, genes specifically expressed by the cells in each cluster.

4.3.2 Graph-Based Clustering

While the clustering algorithms that we have seen in Chapter 3 can be used for single-cell expression data, a different approach is often taken, for example, by the popular analysis suite Seurat [16], where clustering follows the creation of a graph whose nodes are the cells and whose edges join cells with similar transcriptomes.

ℹ K-**nearest neighbor (KNN) graph**

A K-*nearest neighbor* (KNN) graph of the cells assayed in a single-cell transcriptomic assay is built by

(1) computing a suitable distance between all pairs of cells. A possible choice is the Euclidean distance between cells after dimensional reduction obtained by PCA;

(2) placing an edge from each cell to each of its K nearest neighbors

---

Note that the KNN graph is a *directed* graph, since the fact that cell a is among the KNNs of cell b does not imply that b is within the K nearest neighbors of a. Note also that the dimensional reduction used before computing inter-cell distances is *not* the one used for visualization, as it typically reduces the space to $D = 10$ rather than $D = 2$, and uses PCA rather than non-linear methods.

For example, consider the 20 cells represented in Figure 4.5A after dimensional reduction (we will use two dimensions to allow graphical representation, keep in mind that a higher D is used in practice). The KNN graph with $K = 1$ is a directed graph in which each cell is joined to its nearest neighbor, shown in Figure 4.5B, while for higher K, we join each cell with its K nearest neighbors (Figure 4.5C,D).

Figure 4.5 (A) Twenty cells are shown after dimensional reduction to $D = 2$. (B) For $K = 1$, the KNN graph joins each cell to its nearest neighbor (based on Euclidean distance in the dimensionally reduced space). (C, D) For $K = 3, 6$, the KNN graph joins each cell to its three or six nearest neighbors. The graphs are directed, since the relationship 'being among the K nearest neighbor of' is not symmetrical.

A graph naturally produces a clustering of its nodes through the concept of *community*:

> **i** **Community**
>
> A *community* in a graph is a set of nodes that are connected to each other much more than they are connected to nodes outside the community.

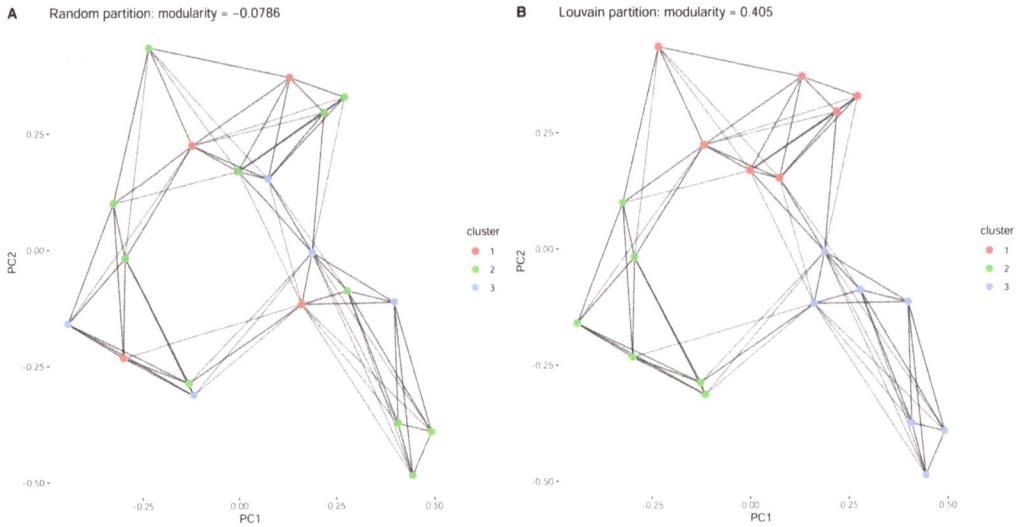

Figure 4.6 (A) The cells are randomly divided into three clusters: The resulting partition has low modularity, i.e. cells are not preferentially connected to cells within the same cluster. (B) The Louvain algorithm finds (heuristically) a partition that maximizes modularity.

A mathematically precise definition uses the concept of *modularity*:

> **i Modularity**
>
> Given a partition of the nodes of the graph into clusters, the *modularity* of the partition is the difference between the fraction of the edges that connect nodes within the same cluster and the same fraction expected from a random partition.

A high modularity implies that the partition of the nodes reflects the true community structure of the graph. Thus, clustering will be performed by looking for the partition of the nodes that maximizes modularity.

Let us start with the KNN graph of our 20 cells, with $K = 6$, and a random partition of the cells into three clusters, represented as colors in Figure 4.6A. To simplify the discussion, we will disregard the directed character of the KNN graph. Since we used a random partition, we expect the modularity to be close to zero, and indeed the edges do not show any special tendency to join nodes of the same cluster: The modularity of this partition is −0.0786. The Louvain algorithm heuristically finds the partition with maximal modularity. In our case the resulting clustering is shown in Figure 4.6B, which, of course, has much greater modularity (equal to 0.405) than the random partition.

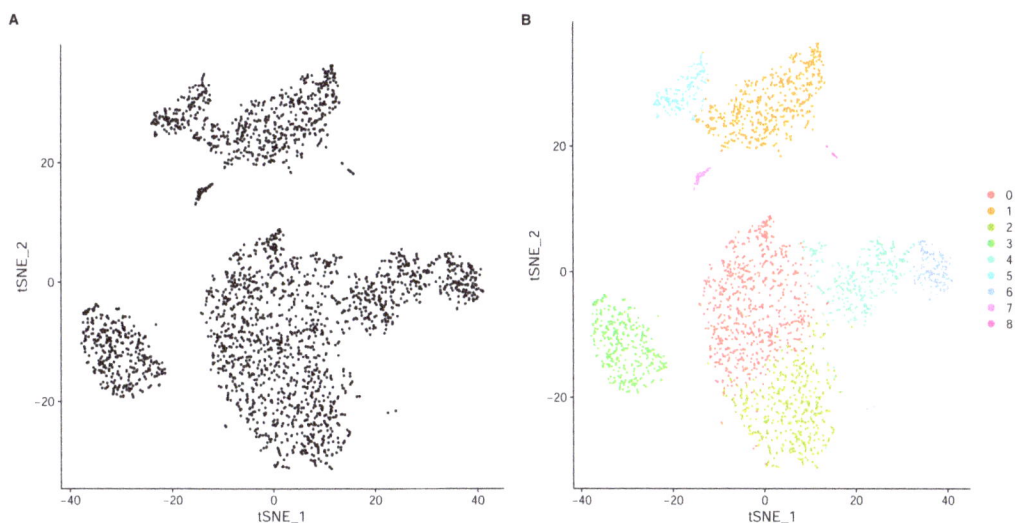

Figure 4.7 (A) Two-dimensional representation of 2,700 peripheral blood mononuclear cells obtained by tSNE. (B) clustering analysis divided the cells into nine clusters, shown as colors.

4.3.3 An Example with Real Data

A typical workflow for the analysis of single-cell RNA-seq data starts by partitioning the cells into clusters using graph-based clustering and representing them, usually colored by cluster, in a plane, using non-linear dimensional reduction. For example, using a dataset of 2,700 peripheral blood mononuclear cells (PBMCs)[2] with Seurat we obtain, using tSNE, the two-dimensional representation shown in Figure 4.7A. Clustering divides the cells into nine clusters, shown as colors in Figure 4.7B. As expected, the cells belonging to each cluster are close together in the tSNE representation, although the clustering was based on a different dimensional reduction (PCA with $D = 10$).

Each cluster thus represents cells sharing a similar transcriptome. The next step is to interpret biologically these clusters in terms of cell types, and to map these cell types into known ones: For example, in PBMCs we expect to find both lymphocytes and monocytes, and to recognize these cell types in the clusters produced by the analysis. This is done through the identification of *markers*, as explained below. However, first a few notes about some aspects of the analysis that were glossed over in our simplified treatment:

1. While the data produced by single-cell RNA-seq are, just like those of bulk RNA-seq, integer counts (number of reads from each cell mapping to each gene), the KNN clustering procedure does not work directly on these counts, but is preceded by filtering of the cells based on various quality parameters, and normalization of the expression values followed by logarithmic transformation.

---

[2]This dataset is used as an example in the Seurat tutorials, and is freely available from 10X Genomics.

2. Seurat actually refines the KNN graph prior to clustering by assigning to each edge joining two cells a weight depending on how many neighbors the two cells share. This graph is called a shared nearest neighbor (SNN) graph and is used for clustering through modularity maximization instead of the original KNN graph.

4.3.4 Cluster Markers

Interpreting the clusters of cells in terms of known or novel cell types is done by first identifying the *markers* of each cluster:

> **i Marker**
>
> A *marker* of a cluster of cells is a gene that is differentially expressed when comparing its expression in the cells belonging to the cluster to that in all other cells.

Both *positive* markers (overexpressed in the cluster of interest) and *negative* ones (underexpressed) are typically extracted. The identification of the clusters is thus a problem of *class comparison*, as discussed in Chapter 2, in which the two classes to be compared are the cells belonging to the cluster of interest and all other cells. Since clusters typically contain hundreds of cells, non-parametric hypothesis testing methods can be used instead of the *t*-test described in Chapter 2. Non-parametric tests have the advantage of not relying on specific assumptions on the distribution of the random variables being measured, such as the assumption of normality underlying the *t*-test. Usually they are less powerful than parametric tests, and thus require large number of replicate measurements. The Wilcoxon test is the most commonly used non-parametric replacement of the *t*-test, and can be used to find cluster markers.

For example, the genes *CD79A* and *MS4A1* are found to be the most significant markers of cluster 3, as shown in Figure 4.8. Both these genes are known to be specifically expressed by B cells: Therefore, we can use these results to posit that cluster 3 represents indeed B cells. By analyzing the list of cluster markers for genes known to be specifically expressed by known cell types we can thus propose a biological interpretation for each cluster.

A popular way of displaying graphically the correspondence between clusters and cell types is to show together the clustering and the expression of a gene specifically expressed by a cell type of interest, as in Figure 4.9, which shows in a visually striking way that the cells classified into cluster 3 are precisely those expressing two genes known to be specifically expressed by B cells.

Note that for each cluster, we define as markers all genes for which the Wilcoxon test shows significant differential expression between the cluster and all other cells. This *P*-value should be considered with some caution, as it is based on the questionable assumption that the individual cells represent *independent* measurements of the expression of the gene. Moreover, in some cases, the top markers by *P*-value are actually expressed in all the clusters, which makes them less useful for interpreting a cluster as a cell type. For example, the top marker of cluster 0 is *RPS6*, with a

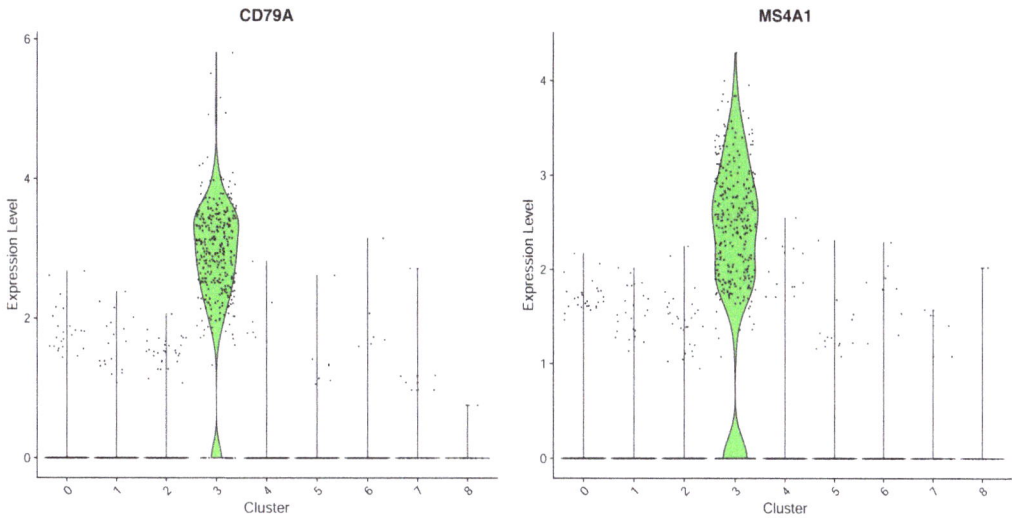

Figure 4.8 Expression of $CD79A$ and $MS4A1$ in the nine clusters. These two genes are markers of cluster 3. Since both are known to be specifically expressed by B cells, we can confidently interpret cluster 3 as mainly containing B cells.

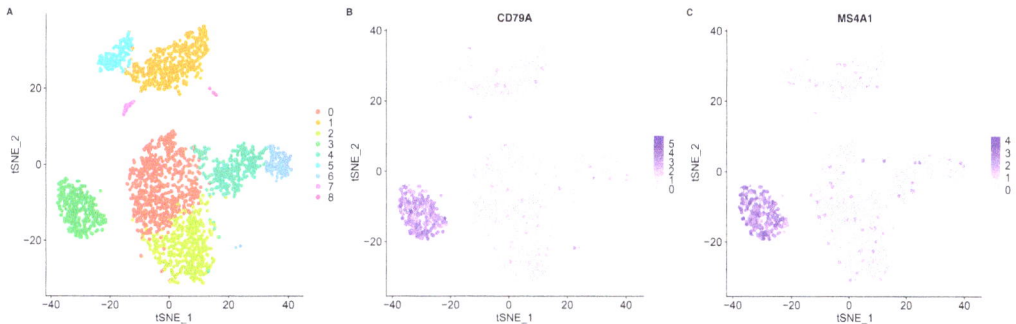

Figure 4.9 The tSNE representation of the cells is shown, in panel A, with colors corresponding to the clusters, and in panels B and C with a color scale representing the expression of $CD79A$ and $MS4A1$, respectively. This representation makes it apparent that cluster 3 (bottom left in panel A) is strongly enriched in cells expressing these two genes, and can thus be interpreted as mainly containing B cells.

Wilcoxon P-value of $5.43 \ 10^{-142}$, whose expression is shown in Figure 4.10. Indeed, RPS6 is a ribosomal protein, expressed by all the cells, but at higher levels in cluster 0. Thus, for purposes of interpretation, the top markers in terms of P-value are not necessarily the most useful ones.

4.3.5 Pseudo-Bulk Analysis

The identification of cluster markers is an example of class comparison applied to single-cell data. Just as in the case of bulk RNA-seq, class comparison can be used

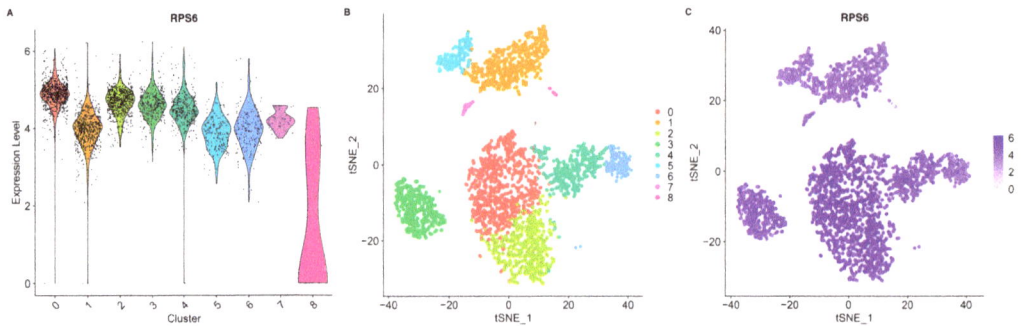

Figure 4.10 RPS6 is the most significant marker of cluster 0 according to the Wilcoxon test. However, as shown by both the violin plot (A) on the left and the tSNE representation (B, C), it is not a satisfactory marker, as it is robustly expressed in all clusters, although at higher levels in cluster 0.

to answer many other biological questions: For example, one could be interested in genes that are differentially expressed between a diseased and a healthy condition in a given cell type (e.g., we might be interested in the differences in gene expression of T cells between individuals infected with a virus and healthy controls). To solve this problem, one could adopt an approach similar to that described for marker analysis, and perform a Wilcoxon test (or a *t*-test) for each gene, comparing its expression in the diseased vs. healthy cells, and limiting the analysis to the cells previously classified into the cell type of interest. However, precisely as in marker analysis, this procedure is likely to produce many false positives because it is based on the hypothesis of independence of gene expression in the individual cells.

The problem can be solved by resorting instead to *pseudo-bulk* analysis:

> **i Pseudo-bulk analysis**
>
> In *pseudo-bulk* analysis the NGS reads assigned to all cells of a given cell type or cluster in a specific biological replicate are aggregated. Differential expression analysis is then performed with the same methods used in class comparison for bulk RNA-sequencing data.

Thus, in our example we would need single-cell RNA-seq data for the T cells of several infected individuals and several healthy controls. After identifying, in each patient, the cell cluster(s) recognizable as made of T cells, we would create pseudo-bulk expression profiles of these cells for each subject, and then find the genes differentially expressed between patients and controls. Thus the subjects, and not the cells, are used as replicates for statistical analysis. Also marker detection can be carried out in this way, as long as biological replicates are available.

4.4 TRAJECTORY INFERENCE

The clusters of cells that we built and used so far are static in nature: There is no notion of a temporal ordering of the clusters. This might seem natural, as most single-cell transcriptomic assays indeed capture and analyze cells that are simultaneously present in a biological sample. However, suppose the sample we are studying is undergoing some dynamical process, such as development, in which precursor cell types differentiate into progressively more specialized ones. This process will not be perfectly synchronous, so that at any given time during the process cells from several differentiation stages will be simultaneously present, and of course understanding which cell types "come first" in the process is crucial if we want to use single-cell transcriptomics to study development. Ideally, we would like to organize all cells into *trajectories* with a well-defined temporal orientation. These trajectories could be linear (from a precursor cell type to a terminally differentiated one through a series of intermediate steps), but also more complex: When a single precursor can generate multiple types of differentiated cells, we expect the trajectory to take the shape of a tree, very similar to those we used in phylogenetic analysis in Chapter 1[3].

Many algorithms have been developed to infer cell trajectories from single-cell transcriptomics. They can be broadly classified into two classes: network-based methods and biology-based ones[4]. The former typically build a network of cells (conceptually similar, and sometimes identical, to the KNN graphs we used for clustering) and use methods of graph theory to reconstruct the trajectory. The latter are based on specific biological assumption on how cell types evolve in time. We will briefly describe one method for each of these classes, our choice being based on the opportunities they present to introduce new useful concepts rather than any judgment on their performance.

4.4.1 Monocle: A Graph-Based Trajectory Inference Method

Also graph-based trajectory inference methods are based on some assumptions about the way cell type mutate into each other during biological processes. The first, quite natural assumption is that changes in the transcriptome are gradual, so that cells that derive directly from each other are more similar in their transcriptome than cells that are separated by many intermediate steps. The second is an assumption about the topology of the trajectories. For simplicity, with the caveats mentioned above, we will consider trajectories described by a tree.

As an example, we will outline the procedure implemented by Monocle [39], one of the most used graph-based trajectory inference tools. The first step taken by Monocle is dimensional reduction, similar to what is usually done before KNN clustering, but performed using a non-linear dimensional reduction algorithm (the specific algorithm

---

[3]In phylogenetic analysis the fact that the trajectory must be a tree is obvious since two species cannot converge back after having diverged. For cells this is much less clear-cut: First, there are dynamical biological processes, such as the cell cycle, in which the trajectory is, obviously, cyclical, and thus cannot be described by a tree. Also in development, cases have been described of transcriptionally distinct progenitors generating differentiated cells that are indistinguishable.

[4]Although also network-based methods are based on some biological assumptions.

used changed with the successive versions of Monocle). The following steps require the introduction of a few new concepts of graph theory:

> **i Weighted graph**
>
> A *weighted graph* is a graph in which a real (usually non-negative) number, the *weight*, is associated to each edge.

The unweighted graphs we have considered so far can be considered as a special case of weighted ones in which every edge has the same weight.

> **i Complete graph**
>
> A *complete graph* is a graph in which every node is connected to every other node by an edge.

Clearly, complete graphs are informative only when they are weighted. Finally,

> **i Spanning tree and minimum spanning tree**
>
> A *spanning tree* of a connected graph is a subset of the edges that forms a tree and connects all the nodes. The *minimum spanning tree* (MST) of a weighted connected graph is the spanning tree with the minimum possible sum of edge weights.

After dimensional reduction, Monocle creates a complete weighted graph with the cells as nodes and their Euclidean distances as weights, then finds the MST of this graph. The MST can have or not have branches. For example, for the 10 cells shown in Figure 4.11A (after dimensional reduction), the MST (Figure 4.11B) has no branches, while for the cells in Figure 4.11C, we obtain a branched MST (Figure 4.11D).

Eventually, the MST will describe the trajectory: To this end, we have to assign a direction in time to each edge. Before we do that, note that the choice of the MST as the trajectory corresponds precisely to the assumption of gradual change of the transcriptome: Biologically, the MST represent the trajectory that minimizes the transcriptomic differences that are adjacent in time.

In the case of an unbranched MST, Monocle interprets the MST as describing a linear trajectory from one end of the MST to the other, and *arbitrarily* chooses one end as corresponding to the beginning of the process and the other as the end. Each cell is assigned a *pseudotime* which increases when going from the beginning to the end. The investigator has to use external, biological knowledge to decide whether the beginning and the end have been chosen correctly or they have to be reversed (this can be done, e.g., by checking the expression of known pluripotency genes, which mark the beginning of the process)[5].

---

[5]The RNA velocity method described below allows establishing the direction of the trajectory in a way that is independent of the specific process under study.

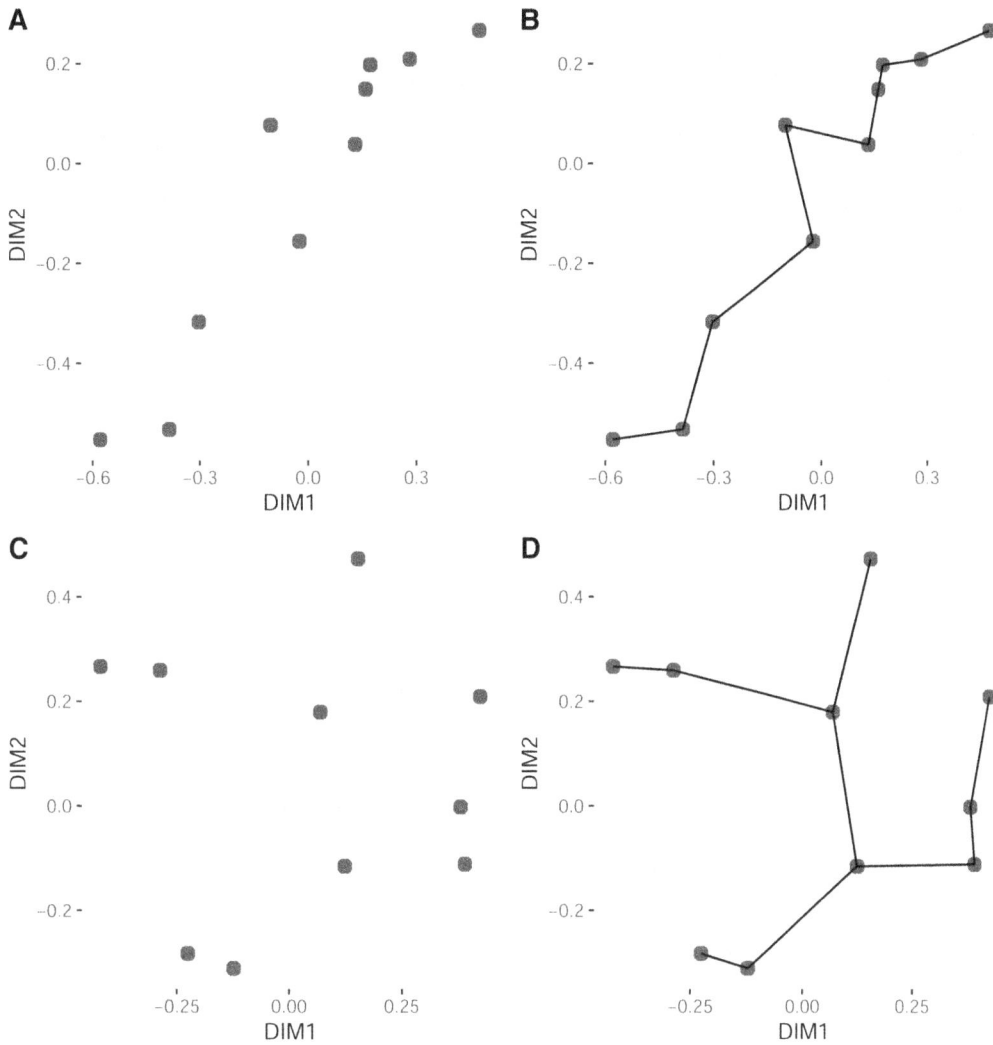

Figure 4.11 Minimum spanning trees of complete weigthed graphs in which each edge is weighted by the Euclidean distance between two cells. (A, B) In this case the MST has no branches. (C, D) an example of branched MST.

The case of a branched MST is dealt with in a more complex way, first by distinguishing between branches due to noise in the data and those of true biological significance. Cells belonging to the latter branches are then assigned a pseudotime value, and hence temporally ordered, with a technically complex procedure which we will not describe in detail.

4.4.2 CytoTrace: A Biology-Based Trajectory Inference Method

CytoTrace [15] can be considered a biology-based trajectory inference method because it introduces a new biological principle in the inference. This principle was established by analyzing a set of scRNA-seq datasets in which the differentiation trajectories were experimentally known, and looking for markers that significantly correlated with differentiation states. A marker is defined here as any quantity that (1) can be assigned to each cell using the scRNA-seq data and (2) significantly correlates with the known differentiation state. Remarkably, a quite simple marker, namely the number of genes detectably expressed by cells (*gene count* in the following) showed strong correlation with differentiation state. Specifically, it was found that cell gene counts *decrease* as cells progress along their differentiation trajectory.

CytoTrace starts by computing the gene count for each cell. According to the principle stated above, this could be directly used as a marker of differentiation state. However, it turns out that better performance can be obtained by (1) identifying the genes whose expression, in the specific dataset, most strongly correlates with gene counts (*top genes*) and (2) using as marker the *gene count signature* (GCS), defined as the average expression of the top genes in each cell[6]. The GCS of each cell is thus used to derive its pseudotime.

4.4.3 RNA Velocity

Finally, we will discuss a method to infer trajectories based on a basic biological fact, namely, that mature, spliced mRNA derives from its unspliced form, which contains intronic sequence. Consider a cell undergoing a change in time of its transcriptome, for example, because it is transitioning from a progenitor to a differentiated state. Since the unspliced RNA is produced before the spliced version, the balance of spliced vs. unspliced RNA of each gene can tell us something about the *future* transcriptome of the cell.

Luckily, it turns out that even if single-cell RNA-sequencing protocol enrich for polyadenylated RNA, the resulting reads still contain a sizable portion of intronic sequence, so that it is actually possible to assess separately the abundance of unspliced and spliced RNA for each (multi-exonic) gene. This is the idea behind the RNA velocity method [21]. Different software packages implement this principle in slightly different ways (corresponding to different assumptions about the differentiation states of the cells present in the sample), but they all produce as their main output an evaluation of the *time derivative* of the expression of each gene in each cell, that is, of how the transcriptome of a given cell is likely to look like in the near future. Roughly speaking, if the future transcriptome of cell A looks like the present transcriptome of cell B, we can conclude that A needs to be placed before B in our trajectory reconstruction.

4.4.4 An Example

As an example, we will consider the data collected in [17], whose authors produced, in particular, single-cell RNA-sequencing of ~2,000 mouse spermatogonic cells. Figure 4.12 shows the results of applying the three methods we have described above to this dataset. Each method assigns a pseudotime value to each cell, while RNA velocity

---

[6]More precisely, the GCS undergoes a smoothing process before being used as a marker.

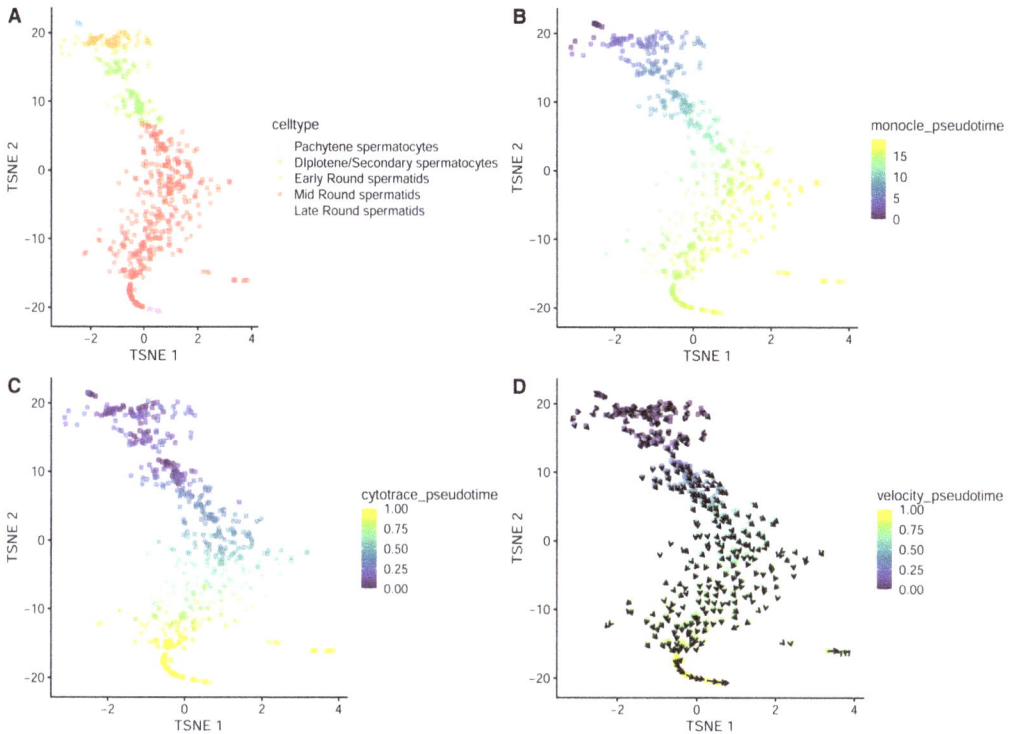

Figure 4.12 (A) tSNE dimensional reduction of the transcriptome of mouse spermatogonic cells. The cells are colored by cluster, and the clusters are interpreted based on the markers they express. (B–D) The same cells colored by pseudotime as assessed by Monocle (B), CytoTrace (C), and RNA velocity (D). The arrows in (D) show the direction of the time derivative of the transcriptome of each cell.

also shows the direction of the time derivative of the gene expression of the cell. It is reassuring that the pseudotime assignments of the three methods are strongly correlated to each other (all correlation coefficients > 0.9), considering that they are based on very different principles. However, this is a rather "simple" situation with cells placed along a single, unbranched differentiation trajectory: The concordance between the methods might not be as high when more complex contexts are examined.

FURTHER READING

Brawand, D., Soumillon, M., Necsulea, A., Julien, P., Csárdi, G., Harrigan, P., Weier, M., Liechti, A., Aximu-Petri, A., Kircher, M., Albert, F., Zeller, U., Khaitovich, P., Grützner, F., Bergmann, S., Nielsen, R., Pääbo, S. & Kaessmann, H. The evolution of gene expression levels in mammalian organs.. *Nature.* **478**, 343–348 (2011). http://dx.doi.org/10.1038/nature10532.

Gulati, G., Sikandar, S., Wesche, D., Manjunath, A., Bharadwaj, A., Berger, M., Ilagan, F., Kuo, A., Hsieh, R., Cai, S., Zabala, M., Scheeren, F., Lobo, N., Qian, D., Yu, F., Dirbas, F., Clarke, M. & Newman, A. Single-cell transcriptional diversity is a hallmark of developmental potential. *Science.* **367**, 405–411 (2020).

Heumos, L., Schaar, A., Lance, C., Litinetskaya, A., Drost, F., Zappia, L., Lücken, M., Strobl, D., Henao, J., Curion, F., Single-cell Best Practices Consortium, Schiller, H. & Theis, F. Best practices for single-cell analysis across modalities. *Nat Rev Genet.* **24**, 550–572 (2023).

La Manno, G., Soldatov, R., Zeisel, A., Braun, E., Hochgerner, H., Petukhov, V., Lidschreiber, K., Kastriti, M., Lönnerberg, P., Furlan, A., Fan, J., Borm, L., Liu, Z., Bruggen, D., Guo, J., He, X., Barker, R., Sundström, E., Castelo-Branco, G., Cramer, P., Adameyko, I., Linnarsson, S. & Kharchenko, P. RNA velocity of single cells. *Nature.* **560**, 494–498 (2018).

Maaten, L. & Hinton, G. Visualizing Data using t-SNE. *J Mach Learn Res.* **9** pp. 2579–2605 (2008).

Ruzicka, W., Mohammadi, S., Fullard, J., Davila-Velderrain, J., Subburaju, S., Tso, D., Hourihan, M., Jiang, S., Lee, H., Bendl, J., PsychENCODE Consortiumğ, Voloudakis, G., Haroutunian, V., Hoffman, G., Roussos, P., Kellis, M. & PsychENCODE Consortium Single-cell multi-cohort dissection of the schizophrenia transcriptome. *Science.* **384**, eadg5136 (2024).

Trapnell, C., Cacchiarelli, D., Grimsby, J., Pokharel, P., Li, S., Morse, M., Lennon, N., Livak, K., Mikkelsen, T. & Rinn, J. The dynamics and regulators of cell fate decisions are revealed by pseudotemporal ordering of single cells. *Nat Biotechnol.* **32**, 381–386 (2014).

CHAPTER 5

Transcriptional Regulation

5.1 INTRODUCTION

In Chapters 2–4, we have seen how the analysis of transcriptomic data can tell us how gene expression changes depending on environmental conditions, tissues, cell types, disease status, etc. In this chapter, we will examine the mechanisms used by the cell to effect gene regulation. As we will see in Chapter 6, these mechanisms are central, in particular, to our understanding of the genetic factors involved in *complex traits*, that is, those phenotypes and diseases that have both genetic and non-genetic causes.

Transcription is the first in a long series of steps leading from the DNA sequence of a gene to the final gene product (a protein or a functional RNA molecule). Thus, the regulation of transcription is the most fundamental mechanism, although certainly not the only one, by which gene expression is orchestrated among tissues or in response to environmental changes.

The transcription of a gene is regulated through the binding of specific proteins, the *transcription factors* (TFs) to the *regulatory regions* of the gene. The binding of most TFs to DNA is *sequence-specific*, meaning that the TF will preferentially bind DNA when a specific *motif* (to be precisely defined below) is present in the sequence. Therefore, changes in the DNA sequence can directly impact gene regulation by changing the affinity of a regulatory region for a TF. However, the binding of a TF is not only determined by the DNA sequence, thus by *genetics*, but also by *epigenetic* modifications of the DNA[1], and in particular by the *state of the chromatin* in the region, which can allow or not the access of TFs to DNA. Moreover, a DNA-bound TF can actually regulate the transcription of a gene only if it is sufficiently close to the transcription start site (TSS) of the gene. This happens when the TF is bound in the vicinity of the TSS (within hundreds to a few thousand base pairs); or when the *three-dimensional conformation of the genome* brings into physical proximity the bound DNA with the region around the TSS, even if they are far away in the genomic sequence.

We will first discuss the analysis of the NGS data that are most useful to study the binding of TFs to DNA, namely, chromatin immunoprecipitation followed by

---

[1]There are many definitions of "epigenetic": Here, we use the term as referring to all modifications of the DNA molecule that do not change the sequence.

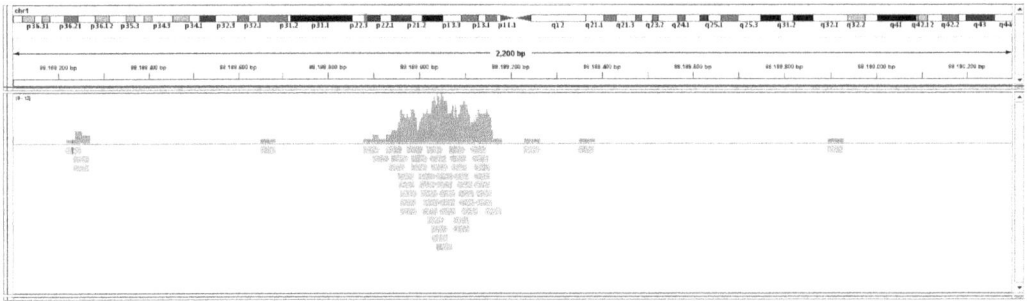

Figure 5.1 ChIP-seq reads aligned to a 2,200 bp region of chromosome 1 and visualized using the integrated genome viewer, showing a typical peak.

sequencing (ChIP-seq). Then we will discuss the genetic aspects of gene regulation, that is how to describe the sequence-specificity of a TF and thus predict its possible binding sites. We will then turn to the epigenetic factors that determine TF binding and its effect on gene expression, including chromatin states and the 3D genome structure. In terms of analytical tools, we will re-encounter some mathematical objects introduced in Chapter 2, such as the Poisson distribution and likelihood ratios; we will describe general methods to evaluate the performance of a *classifier*, including *receiver operating characteristic (ROC) curves*; and we will introduce *Markov chains* and *hidden Markov models*.

5.2   TRANSCRIPTION FACTOR BINDING AND MOTIF ANALYSIS

5.2.1   ChIP-seq

Our main tool to study TF binding at the scale of the whole genome is Chromatin ImmunoPrecipitation followed by sequencing (ChIP-seq). Like RNA-seq and Hi-C (described below), ChIP-seq is an example of how NGS techniques revolutionized the whole field of molecular biology, and not only genetics (defined as the study of the DNA sequence). ChIP-seq uses an antibody specific for the TF of interest to immunoprecipitate the chromatin and obtain a sample of DNA enriched in those regions that were originally bound by the TF. Sequencing of this DNA and alignment to the reference genome allow us to catalog the genomic loci that were bound by the TF (TF binding sites - TFBSs) and hence to hypothesize the genes whose expression is regulated by the TF. Here, we will discuss how these data are analyzed to produce a list of binding sites through the *peak calling* procedure, and how the results of peak calling can be used to infer the sequence specificity of the TF.

5.2.2   Peak Calling

The reads produced by a ChIP-seq experiment are first aligned to the reference genome, as discussed in Chapter 1. Figure 5.1 shows the alignments obtained from a ChIP-seq experiment, carried out in human embryonic stem cells, on the pluripotency-related TF POU5F1 in a 2,200 bp region of chromosome 1.

The distribution of the aligned reads is precisely what we would expect after the enrichment produced by ChIP: Far from being uniformly distributed across the

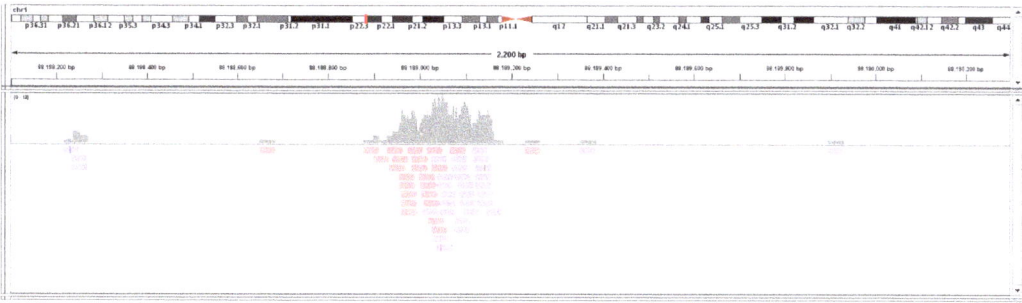

Figure 5.2 The aligned reads shown in Figure 5.1 are colored based on the genomic strand to which they align. The separation is due to the fact that reads coming from each strand will be shifted in opposite directions with respect to the original fragment.

genome, the reads accumulate in localized regions, whose size is typically of the order of hundreds of base pairs, corresponding to the genomic loci originally bound by the TF. Given their appearance when visualized in a genome browser such as the IGV, these regions are called *peaks*.

> **i Peak**
>
> A *peak* is a genomic region showing significantly more aligned reads than the surrounding genome.

The analytical procedure allowing us to detect peaks in the genome is known as *peak calling*. We will describe one of the most commonly used algorithms for peak calling, MACS, introduced in [41].

Let us look again at the peak shown above, but this time let us distinguish by color the reads aligning to the two strands of the genome (Figure 5.2). It is easy to see that these are quite separated: This is due to the fact that DNA fragments produced by ChIP are typically longer than the NGS reads. Since the reads represent the 5' end of each fragment and come from both strands, reads coming from each strand will be shifted in opposite directions with respect to the original fragment.

To compensate for this, MACS first uses high-quality peaks, detected by a preliminary analysis, to determine the distance d between reads coming from the two strands. Then, all reads are shifted by $d/2$ in a direction depending on the strand, so that after the shift the reads are centered in the most probable location of the DNA/protein interaction. Peak calling is performed on these shifted reads.

To state the problem of peak calling in quantitative terms, we want to find the genomic regions in which the density of aligned reads is much higher than in the rest of the genome. For example, suppose we count 50 reads in a given 300 bp region, while the genome-wide average number of reads in a 300 bp region is equal to 5: We then have a *10-fold enrichment* of reads in our region compared to the rest of the genome.

As usual, we need to ask whether this enrichment could have happened by chance, and the right way to answer this question is to formulate and test a null hypothesis.

The null hypothesis is that the reads are randomly distributed over the genome. As the data come from next-generation sequencing, they are, just like RNA-seq data, digital in nature, so that the Poisson distribution is a suitable null hypothesis. We need to specify the λ parameter of the Poisson distribution, and the simplest way to do that is to choose the genome-wide average number of reads in a region. Thus, in the example described above, we would have $\lambda = 5$ and the probability of finding k reads in a 300-bp region in the null hypothesis would be

$$P(k; \lambda) = \frac{\lambda^k e^{-\lambda}}{k!} = \frac{5^k e^{-5}}{k!}$$

shown in Figure 5.3, from which we learn that the actually observed 50 reads are a very unlikely value under the null hypothesis. Specifically, the P-value is the probability of finding 50 or more peaks in a region and is given by

$$\sum_{k=50}^{\infty} P(k; 5) = 2.18 \ 10^{-32}$$

Thus our region can be called a peak, and we have high confidence that the TF was actually bound to this genomic locus.

To systematically find all peaks, we need to choose the size of the regions to be tested, and then apply the same procedure to the whole genome. MACS uses $2d$ as the region size, that is, twice the distance between the reads coming from the two strands determined above, which is an estimate of the typical size of the peaks. Moreover, it turns out that simply choosing λ to be the number of reads in a region of size $2d$, averaged over the whole genome, would be too simplistic, as it is known that different regions of the genome have different probabilities of generating reads, irrespective of any ChIP enrichment. Thus, MACS uses a *local* λ computed in a region around the window of interest, and specifically:

$$\lambda = \max(\lambda_{BG}, \lambda_{5k}, \lambda_{10k})$$

where λ_{BG} is computed over the whole genome and λ_{5k}, λ_{10k} are computed on 5,000 and 10,000 bps centered on the region of interest. Using the max is a *conservative* choice: To call a peak, we need the read count to be significantly higher than the genome-wide average *and* the local 10 Kbp average *and* the local 5 Kbp average.

In many cases, besides the ChIP sample, also a control DNA sample is sequenced, for example, the DNA without immunoprecipitation or immunoprecipitated with a non-specific antibody. In this case the control sample, instead of the ChIP sample, is used by MACS to compute λ:

$$\lambda = \max(\lambda_{BG}, \lambda_{1k}, \lambda_{5k}, \lambda_{10k})$$

where all λs are now computed on the control sample and the average within 1 Kb has been added (1 Kb cannot be used when using the ChIP sample because 1 Kb is not larger enough than the typical peak size).

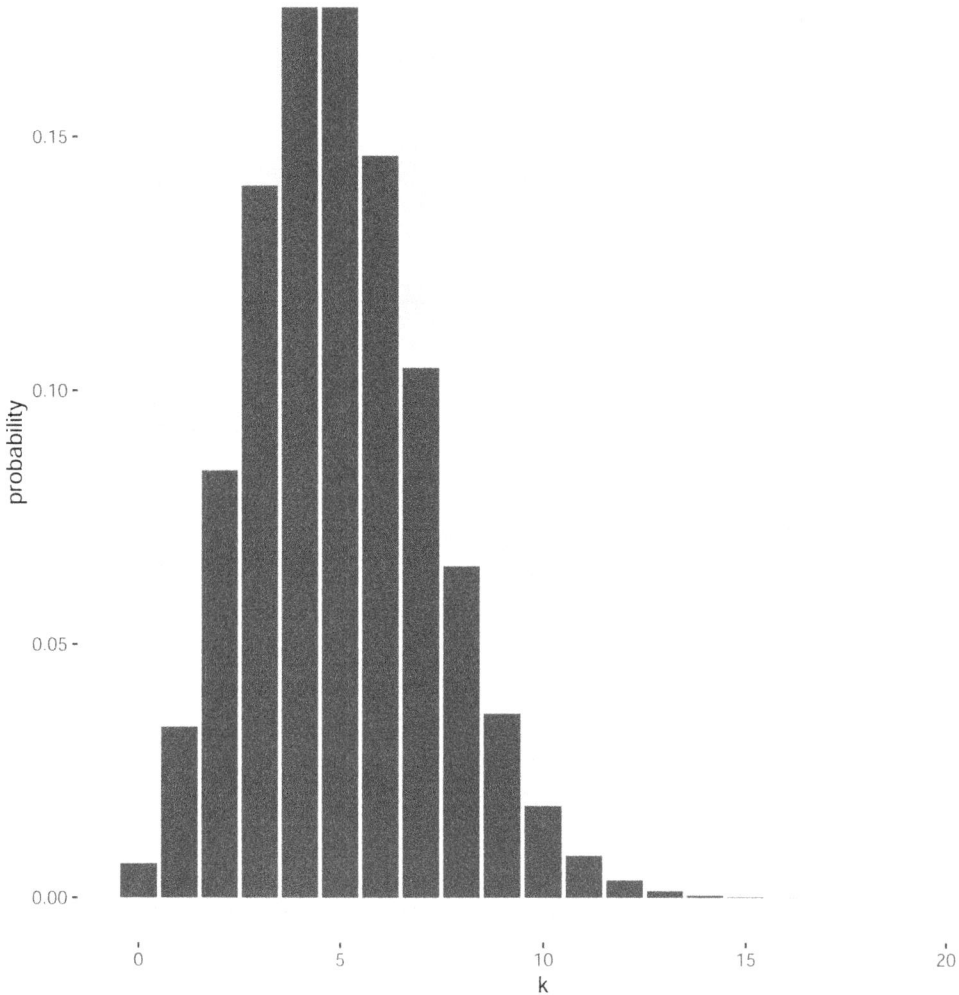

Figure 5.3 Probability of finding k reads aligned to a 300 bp genomic region in the null hypothesis in which reads are uniformly distributed on the genome, and the average number of reads aligned to a 300 bp genomic region is $\lambda = 5$.

After this procedure, the results of a ChIP-seq experiment are expressed as a list of peaks, and can be visualized in a genome browser, where often the read distribution is shown together with the peak location. Figure 5.4 shows an example of a region near the TSS of the *NANOG* gene with a peak derived from a ChIP-seq experiment on the POU5F1 TF. These are two TFs involved in pluripotency and known to regulate each other; the gray bars represent the peaks found by the statistical analysis and the green tracks represent the reads. Note that the experiment was conducted in replicate and the peak appears in both replicates.

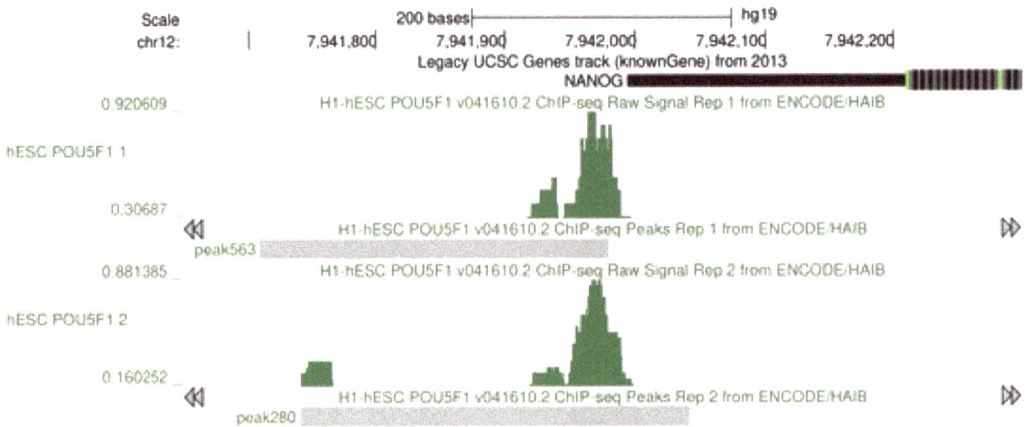

Figure 5.4 A genomic region near the transcription start site (TSS) of *NANOG*, a transcription factor involved in pluripotency, is shown in the UCSC genome browser. The two 'hESC POU5F1' tracks show the peaks from two replicate ChIP-seq assay for the binding sites of POU5F1, another transcription factor involved in pluripotency. The peaks are represented by gray bars, while the green histograms represent the density of aligned reads. The fact that two replicate experiments produce almost identical peaks increases our confidence in the result. Since the peaks are located close to the *NANOG* TSS, they strongly suggest that POU5F1 is involved in the transcriptional regulation of *NANOG*.

Once we have the peaks (or, more generally, a collection of empirically determined binding sites of a TF), we can use them to understand the sequence specificity of the TF. If the TF recognizes a specific sequence, such sequence should be found in all or at least most of the peaks. In the next subsection, we will see how this analysis is performed to produce a description of the sequence specificity of a TF, expressed by a *positional frequency matrix*.

5.2.3 Positional Frequency Matrices

A straightforward way to investigate the sequence specificity of a TF is to obtain the sequence of many bound genomic loci, for example, from a ChIP-seq assay, and use specific algorithms for multiple local alignment to find regions that are highly similar among them. Since the typical sequence recognized by a TF is rather short (typically 4-15 bps) while the typical ChIP-seq peak size (i.e. the ChIP-seq *resolution*) is of the order of a few hundred bps, only a small portion of each peak sequence will produce a high-scoring alignment. This portion is isolated from the alignment and used to compute a *positional frequency matrix* (PFM), which counts how many times each base was found in each position of the alignment. For example, in the case of POU5F1, the multiple local alignment identifies a region of length 11 bps, ten aligned sequences being shown below:

1	2	3	4	5	6	7	8	9	10	11
C	G	A	G	G	C	A	A	A	C	C
T	T	T	T	G	C	A	A	A	C	G
A	A	A	T	G	C	A	A	A	A	A
A	T	A	T	G	C	A	A	A	A	A
T	T	A	T	G	C	T	A	A	T	G
A	A	A	T	G	C	A	A	A	C	C
G	C	A	T	G	C	T	A	A	T	G
G	C	A	T	G	C	A	A	A	A	A
T	A	A	T	G	T	A	A	A	A	A
A	A	A	T	G	G	A	A	A	T	A

In these sequences, positions 5, 8, and 9 always contain, respectively, G, A, and A, while other positions are variable. This suggests that POU5F1 binds an 11-bp motif, with some positions strongly constrained to contain a given base, and others (such as positions 1, 2, 10, and 11) which can contain essentially any base.

> **i Positional frequency matrix**
>
> A *positional frequency matrix* describes the sequence specificity of a TF by reporting, for each position in the sequence bound by the TF, how many times each base was found in a collection of empirically determined binding sites.

For the ten sequences shown above, the PFM would thus be

	1	2	3	4	5	6	7	8	9	10	11
A	4	4	9	0	0	0	8	10	10	4	5
C	1	2	0	0	0	8	0	0	0	3	2
G	2	1	0	1	10	1	0	0	0	0	3
T	3	3	1	9	0	1	2	0	0	3	0

while the following is the PFM reported in the JASPAR database, derived from 13,114 ChIP-seq peaks

	1	2	3	4	5	6	7	8	9	10	11
A	4982	3603	12693	83	717	533	11756	10560	12280	2556	3476
C	1290	2785	122	130	192	10310	132	690	270	2104	2450
G	1996	1908	229	67	11571	548	118	1026	257	2071	3551
T	4846	4818	70	12834	634	1723	1108	838	307	6383	3637

For each position, we can compute an *information content* (IC), measured in bits, that expresses how much we know about which base(s) can be in that position for

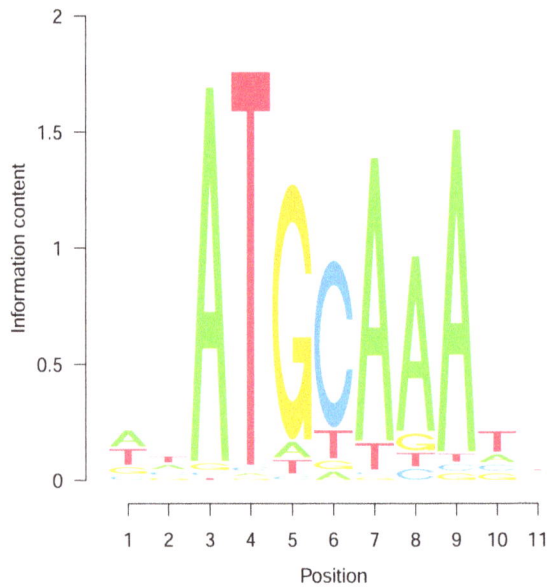

Figure 5.5 Logo representing the positional frequency matrix of TF POU5F1 derived from the alignment of 13,114 peaks. The height of each column is proportional to the information content, in bits, of the corresponding position. Within each position, the height of each letter is proportional to its frequency.

binding to occur. For example, for a position containing always the same base the information content is equal to 2 bits (we know precisely which one base of four needs to be in this position, and 2 bits are required to specify an object of four possible choices). Conversely, a position for which each base is equally likely would have an IC of zero. A general formula allows computing the IC for every distribution of the four frequencies. For our POU5F1 PFM, the IC of each position is represented in the PFM logo (Figure 5.5), which also shows, for each position, the possible letters, their relative height being proportional to their frequency.

5.2.4 Scoring Candidate Binding Sites

A PFM for a TF describes its sequence-specificity, and thus can be used to *predict* binding sites on the genome. This is useful even if we can use ChIP-seq to measure the actual binding, for at least three reasons:

1. First, ChIP-seq measures binding in the specific cellular context in which the assay was performed, and the same TF can bind different sites in different cellular contexts, for example, due to differences in chromatin accessibility (see below in this chapter).

2. If we can predict binding, we can, in particular, predict the effect of a DNA variant on binding, and thus on gene regulation. For example, we can predict that a mutation in position four of the POU5F1 binding site would severely

impact the binding, and thus gene regulation, even if we do not have a direct measurement of the binding on the mutated sequence. This is especially important in the interpretation of non-coding variants associated to diseases, as we will see in Chapter 6.

3. Finally, TFBSs can be predicted from PFMs in species for which we have the genomic sequence but not ChIP-seq experiments. This is possible because the TFs themselves, and in particular their DNA-binding domains that determine sequence specificity, are often strongly conserved between species also at very large evolutionary distances[2]. On the contrary, the set of genes targeted by a TF tends to evolve quite rapidly through changes in the gene regulatory sequences, so that being able to predict binding from sequence allows us to apply comparative genomics to the study of the evolution of gene regulation.

Intuitively, it is quite obvious how we can use a PFM to predict whether a TF will bind a DNA sequence located in a regulatory region: For example, for POU5F1, a look at the PFM logo suggests that the sequence ATATGCAAATA will have a high probability of being bound, while ATCCATCCCGA will not. To make this observation quantitative, we first transform our PFM into a *positional probability matrix* (PPM) by transforming frequencies into probabilities (i.e., dividing each entry by the sum of the corresponding column). Thus our PPM is:

	1	2	3	4	5	6	7	8	9	10	11
A	0.38	0.275	0.968	0.006	0.055	0.041	0.896	0.805	0.936	0.195	0.265
C	0.098	0.212	0.009	0.01	0.015	0.786	0.01	0.053	0.021	0.16	0.187
G	0.152	0.145	0.017	0.005	0.882	0.042	0.009	0.078	0.02	0.158	0.271
T	0.37	0.367	0.005	0.979	0.048	0.131	0.084	0.064	0.023	0.487	0.277

Then, we use an approach that is frequently used in the analysis of biological sequences, namely the use of *generative models* compared through likelihood calculations.

ℹ **Positional probability matrix**

A *positional probability matrix* is derived from a PFM by transforming the frequencies into probabilities, and can be used as a generative model capable of producing binding sites, and thus to score genomic sequences for their ability to bind the TF through likelihood comparison.

---

[2]A classical example is that of PAX6, a TF involved, in particular, in the development of the vertebrate eye. This TF has an orthologous TF in *Drosophila melanogaster* called *eyeless* (from the phenotype of the flies with mutations in this gene). However, the ectopic expression of the mouse Pax6 is able to drive the formation of ectopic eyes in *Drosophila* (although the vertebrate and insect eyes are structurally very different). Thus this TF has been conserved since the common ancestor of bilaterians, and was probably initially involved in the development of some primitive form of eye, or photosensitive organ, which then separately evolved into the vertebrate and insect eyes.

First, we interpret our PPM as a model that can generate binding sites of POU5F1 as sequences of length 11. In these sequences, each base appears in each position with the probability observed in experimentally determined POU5F1 binding sites, and collected in the PPM. This generative model will be used to determine whether a given sequence that we observe (e.g., in the regulatory region of a gene) is likely to be able to bind POU5F1, through a process involving the concept of likelihood that we have previously seen in Chapter 2.

Specifically, we define another generative model, the *background* model, which describes not the POU5F1 binding sites, but how sequences of length 11 are distributed in the whole genome. We then ask whether the sequence observed is more likely to have been generated by the PPM or the background. To make things simple, let us describe the background as a purely random DNA sequence[3]: Thus, in every position of the 11 generated by the background model, each of the four bases can appear with probability $1/4$.

Let us now consider, for example, the sequence $S_1 = ATATGCAAATA$, which from the PFM logo seems to be a good candidate to be a POU5F1 binding site. As introduced in Chapter 2, the *likelihood* of a model given some data is the probability that the model would generate the data. In this case the data is the observed sequence $ATATGCAAATA$ and we want to compare the PPM with the background model. The likelihood of the PPM model is thus:

$$L(PPM|S_1) = \prod_{k=1}^{11} PPM_{s(k),k} = 0.38 \cdot 0.367 \cdot 0.968 \cdots = 0.00798$$

where $PPM_{s,j}$ ($s = A, C, G, T; j = 1 \ldots 11$) is the entry in row s and column j of the PPM, and $s(k)$ is the letter found at position k in our sequence. The likelihood of the background model is simply:

$$L(background|S_1) = \left(\frac{1}{4}\right)^{11} = 2.38 \ 10^{-7}$$

and is thus much smaller than $L(PPM|S_1)$. Thus, we conclude that the sequence S_1 is much more likely to have been generated by the PPM than by the background, and is thus a good candidate binding site for POU5F1.

The same calculations for $S_2 = ATCCATCCCGA$ give:

$$L(PPM|S_2) = 4.21 \ 10^{-14}$$

while $L(backround|S_2)$ is the same as $L(backround|S_1)$, so that in this case we conclude that S_2 is much more likely to have been generated by the background model than by the PPM, and thus is not a good candidate binding site for POU5F1.

More generally, given a sequence S and a PPM, we define the *score* of S as the $\log_2$ of the ratio of the likelihoods of the PPM and the background model:

$$score(S_1) = \log_2 \frac{0.00798}{2.38 \ 10^{-7}} = 15.03$$

---

[3]This is not an accurate description of how bases are distributed in the whole genome. A better background model would reflect the actual base composition of the genome, or better of the known regulatory regions.

$$score(S_2) = \log_2 \frac{4.21 \ 10^{-14}}{2.38 \ 10^{-7}} = -22.43$$

In practice, the calculations are never performed directly in terms if likelihoods, but rather of their logarithms, because the likelihoods are often very small numbers that might create computational problems[4]. A third type of matrix, called a *positional weight matrix* (PWM) is defined that reports in each slot the $\log_2$ of the PPM minus the $\log_2$ of the background probability of the corresponding base in the background, so that the score of a sequence is simply:

$$score(S) = \sum_k PWM_{s(k),k}$$

To avoid computational problems due to zeros in the PFM (which would give $-\infty$ in the PWM), a small number called a *pseudocount* is added to all elements of the PFM prior to transforming it into a PPM and then into a PWM.

The score we assign to a candidate binding site is thus a likelihood ratio (LR). While in class comparison for RNA-seq data, we used the LR to compute a *P*-value, here this cannot be done, since the two models being compared are not nested, and hence the log of the LR cannot be assumed to follow a χ^2 distribution (as hinted by the fact that here the log of the LR can be positive or negative). Several packages for the identification of TFBSs from PPMs associate a *P*-value to the score using empirical procedures.

5.2.5   Including the Reverse Complement Sequence

Since the DNA molecule to which TFs bind is double-stranded, when we evaluate the binding potential of a regulatory sequence we actually need to compare *two* sequences and evaluate their score: the sequence as reported in the reference genome, and its *reverse complement*. The latter is obtained by reversing the sequence and replacing each base with its Watson-Crick paired base, and is thus the sequence read on the other DNA strand, going from 5' to 3'. For example, the reverse complement of GATTACA is TGTAATC:

```
5' - GATTACA - 3'
     |||||||
3' - CTAATGT - 5'
```

The TF will bind either strand if it finds its motif, although its orientation will depend on which strand is bound. In general, the orientation of a TF does not significantly impact its regulatory function, and therefore the analysis described above is performed on both strands: The score assigned to a sequence is actually the maximum between the score of the sequence and that of its reverse complement. An important exception is the role of transcription factor CTCF in three-dimensional genome organization, which we will discuss below.

---

[4]Since computers use a finite number of bits to represent a real number, the representation is actually discrete, and very small numbers cannot be distinguished from zero – how small depends on the number of bits used.

5.2.6 Predicting Binding Sites

Let us now use this machinery to take a sequence of interest (e.g., the *promoter* of a gene, i.e., the regulatory region close to its transcription start site) and predict whether it will bind a TF for which we have a PWM. Based on the discussion above, a natural choice is to predict that a region will bind the TF if it contains at least one subsequence of length 11 with score > 0.

Let us consider a high-confidence set of sites experimentally known to bind POU5F1, namely, the intersection of the peaks found in the two replicate ChIP-seq experiments in human embryonic stem cells described above (1,225 sequences, considering only those of length > 100 bp). For each peak sequence, let us compute the score of all its subsequences of length 11. As expected, most peak sequences (1,223) do contain at least one site with score > 0.

However, as a negative control, we can build a set of sequences containing as many sequences as the POU5F1 peaks, of the same length, but chosen randomly on the genome, and such that they do not overlap any peak. Presumably these sequences do not bind POU5F1, but if we analyze them with the PWM we still get 1,096 sequences[5] with at least one site with score > 0. Thus, predicting a sequence to be able to bind POU5F1 based on the existence of a site with score > 0 is not very effective, and in particular lacks *specificity*.

To discuss these issues a bit more formally, consider the set of sequences consisting of the peaks and the randomized sequences. As the former sequences are able to bind POU5F1, we will call them *positives*, while the randomized sequences, which (presumably) cannot bind POU5F1 are called *negatives*. Given a *classifier* claiming to predict which sequences will bind POU5F1, we can compute four numbers:

- TP (true positives): number of positives that are predicted to be positive
- FP (false positives): number of negatives predicted to be positive
- TN (true negatives): number of negatives predicted to be negative
- FN (false negatives): number of positives predicted to be negative

For our classifier we have:
$$TP = 1223$$
$$FP = 1096$$
$$TN = 1225 - 1096 = 129$$
$$FN = 1225 - 1223 = 2$$

These number can be used to compute *accuracy*, *specificity*, and *sensitivity*:
$$acc = \frac{TP + TN}{P + N} = 0.552$$

---

[5]To clarify: This is the number obtained from a specific set of random sequences built as described. Obviously each time this procedure is repeated it will produce a slightly different number of sequences predicted to bind the TF.

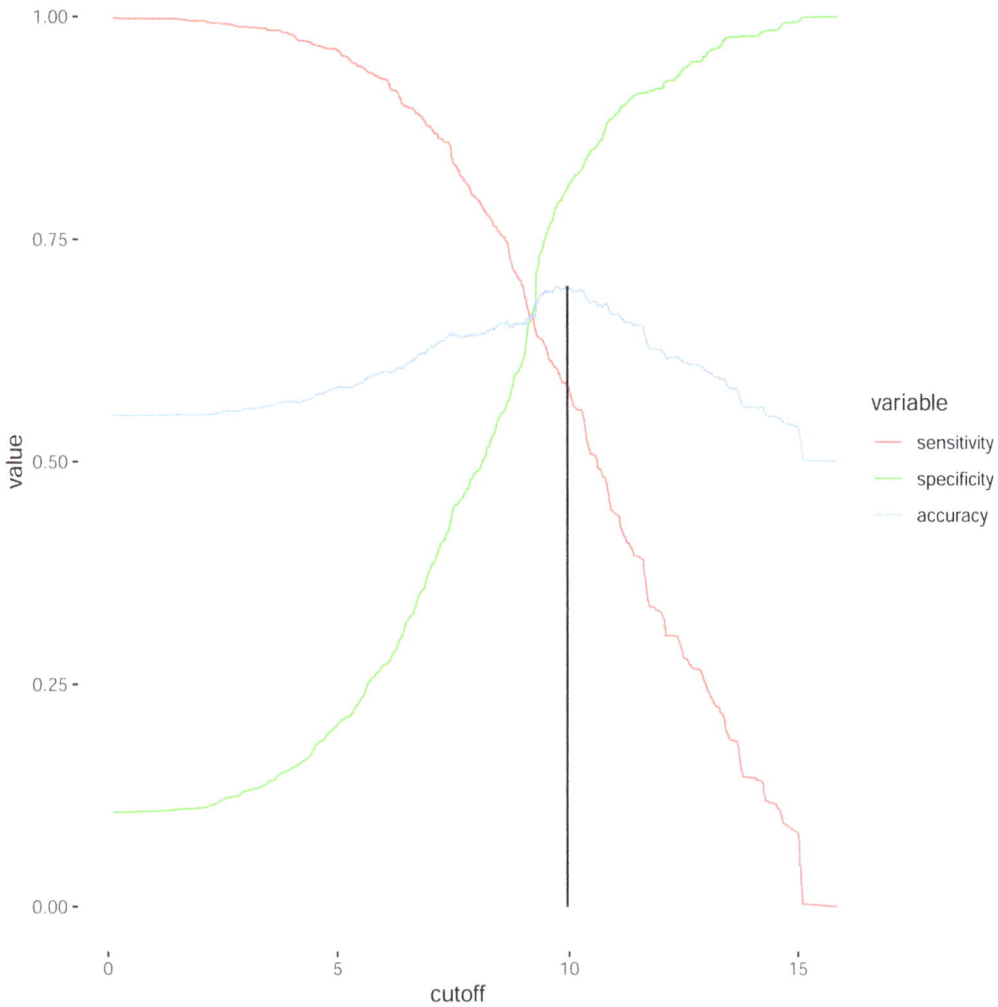

Figure 5.6 Performance of a classifier which predicts a sequence to be bound by POU5F1 if it has at least one site scoring better than a cutoff, as a function of the cutoff. The vertical line corresponds to the maximum accuracy.

$$sens = \frac{TP}{TP + FN} = \frac{TP}{P} = 0.998$$

$$spec = \frac{TN}{TN + FP} = \frac{TN}{N} = 0.105$$

where P and N are the total number of positive and negative sequences. A perfect classifier would have accuracy, sensitivity, and specificity equal to one; high accuracy indicates that the predictions of the classifier tend to be correct; high specificity indicates that the classifier tends to correctly identify negatives; and high sensitivity that it tends to correctly identify positives. Thus our classifier has very high sensitivity but very low specificity, as it tends to predict almost all sequences as positives. Therefore, its accuracy is very close to what we would obtain by tossing a coin.

Presumably specificity can be improved by setting a higher cutoff on the score, that is, by predicting that sequences with at least one site scoring $>C$, with $C > 0$, will bind the TF. If we do this systematically for increasing values of C, we get the behavior of our three performance measures shown in Figure 5.6. At the two ends of the cutoff range, the accuracy is close to 0.5, that is, the same accuracy that we would obtain by tossing a coin. It shows a maximum of 0.697 at the cutoff value 9.96 (showed by the black vertical segment). This might be a reasonable choice of cutoff, and corresponds to a specificity of 0.808 and a sensitivity of 0.586.

A global evaluation of the performance of a classifier, which does not require the choice of a specific cutoff, can be obtained with the *receiver operating characteristic (ROC) curve*[6]. Since specificity (sensitivity) is an increasing (decreasing) function of the cutoff, it is possible to ditch the cutoff and plot one as a function of the other.

> **i Receiver operating characteristic curve**
>
> Given a classifier depending on a cutoff, the *receiver operating characteristic (ROC) curve* represents the sensitivity of the classifier as a function of its specificity.

Usually what is actually plotted is sensitivity as a function of 1-specificity. For our classifier we get the red curve shown in Figure 5.7. An ROC curve always goes from the origin (specificity = 1, sensitivity = 0: all cases predicted to be negatives) to the point of coordinates (1,1) (specificity = 0, sensitivity = 1: all cases predicted to be positives), and gives the sensitivity of the classifier for any given value of the specificity. For example, in our case, the curve tells us that if we choose the cutoff so as to have a specificity of 0.5, the sensitivity will be 0.787 (vertical line). The dotted line represents the performance of a random classifier, that is, of a score completely unrelated to the response we want to predict.

The global performance of a classifier can be assessed, without having to choose a specific cutoff, by computing the *area under the curve* (AUC). The AUC has a simple probabilistic interpretation, namely, it is the probability that randomly choosing a positive and a negative sequence, the classifier will have a higher value for the positive than the negative sequence. For our classifier the AUC is 0.744, which is better than the AUC of a random classifier (AUC = 0.5), but quite far from that of a perfect classifier (AUC = 1, since a perfect classifier will always give a higher score to a positive than to a negative sequence).

Thus, whether you judge it by its maximum accuracy of 0.697 or by the AUC of 0.744, using the score of the best predicted binding site as a classifier of actual POU5F1 binding is not very satisfactory. A few improvements can be made to PWM-based classifiers, such as requiring not simply the presence of a high-scoring site but a statistical enrichment of such sites; or requiring the *evolutionary conservation* of the

---

[6]The curious name derives from the fact that the technique was originally developed during World War II to evaluate the performance of military radar receivers.

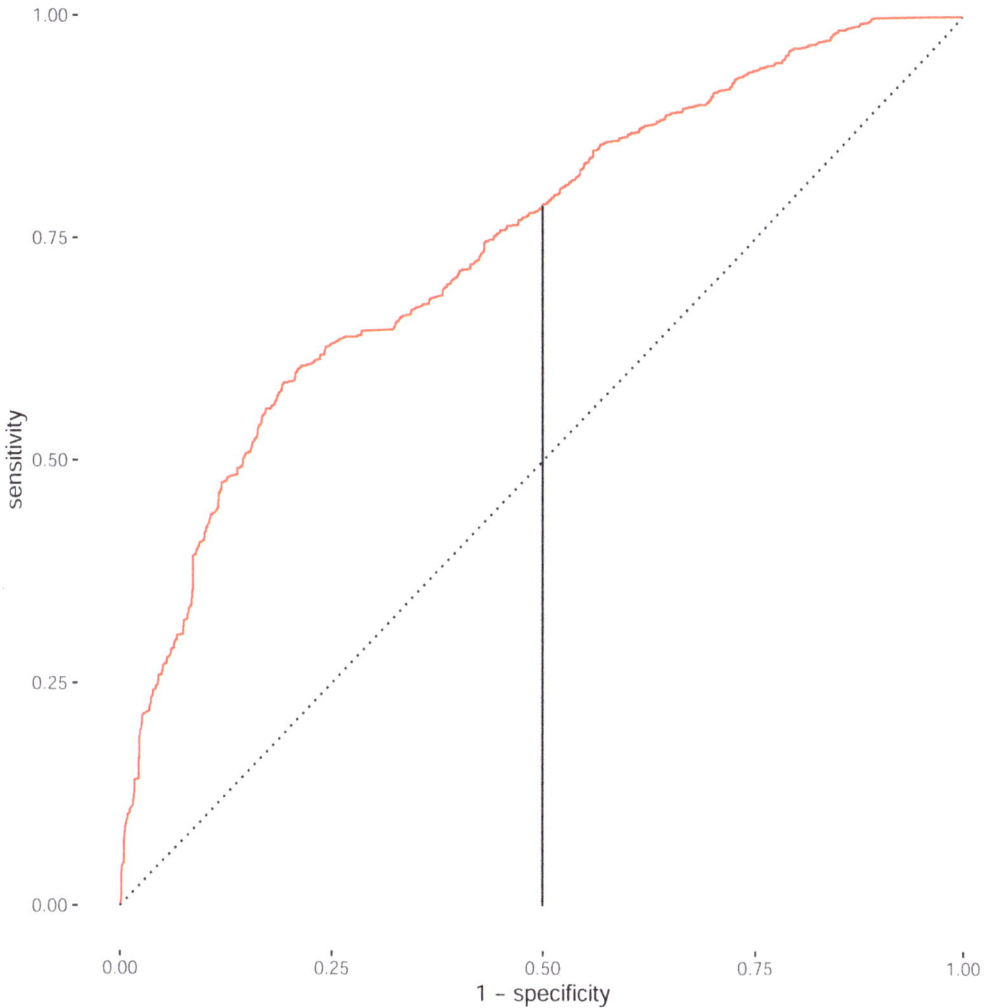

Figure 5.7 The ROC curve for our predictor of POU5F1 binding sites (red line). The curve allows computing the sensitivity corresponding to any given specificity value (or vice versa). For example (black vertical line) if the cutoff is chosen as to give a 50% specificity, the sensitivity is 0.787. The dotted line shows the ROC curve expected of a random predictor.

binding site, that is, its presence in the orthologous region of other species, since this suggests negative selective pressure and hence functional relevance, as discussed in Chapter 1. More sophisticated classifiers based on *machine learning* will be discussed in Chapter 6.

In general, it is fair to say that the prediction of TF binding from sequence data alone is a very challenging task. This is mainly due to the fact that, *in vivo*, the binding of TFs depends on many factors beside the sequence itself, including in particular the *chromatin states* discussed below.

5.3   CHROMATIN STATES

5.3.1   Open Chromatin and ATAC-seq

In the simplest classification scheme, the chromatin of a genomic region can be in an *open* or a *closed* state. Open chromatin is *accessible*, meaning in particular that TFs can physically interact with DNA and bind it if they find the appropriate sequence motif. Closed chromatin, on the contrary, is not accessible to TFs, so that even if a closed chromatin region contains the motif recognized by a TF, it does not affect gene expression as the binding cannot take place.

Therefore, to understand transcriptional regulation, it is not enough to determine whether a regulatory region contains the motif recognized by a TF, but it is also necessary to determine whether, in the context of interest (tissue or cell type), the region is in an active, open state. Several sequencing-based experimental assays have been developed to asses open vs. closed chromatin, including DNase I hypersensitive sites sequencing (DNase-seq) and formaldehyde-assisted isolation of regulatory elements (FAIRE-seq). The most recently developed method, known as ATAC-seq (assay for transposase-accessible chromatin using sequencing) has rapidly become the most popular, also because it can be readily applied also at the single-cell level.

Technically, ATAC-seq is based on the activity of a mutated, hyperactive transposase (Tn5) that selectively cleaves DNA in open chromatin regions. The fragments obtained are sequenced and aligned to the reference genome: Thus the regions that were in an open chromatin state appear as enriched in read density, similarly to what we saw for ChIP-seq. Indeed, the first step of the analysis of ATAC-seq data is precisely the same as that of ChIP-seq: Regions of enrichment are determined by peak-calling, for example, using the MACS2 software described above. The resulting peaks provide a catalog of all the genomic regions that are accessible in the context of interest, and are strongly correlated with the transcriptome, since open chromatin regions tend to be close to expressed genes.

Besides this basic analysis, ATAC-seq data can be used to ask more complex questions. For example, we can investigate *differential accessibility* by finding the regions with significantly different enrichment of ATAC-seq reads between two biological conditions. This is done with methods very similar to those used for class comparison in RNA-sequencing data: Also in this case we are interested in statistically significant differences in the number of reads aligned to a region between two conditions.

5.3.2   Histone Modifications and Chromatin States

While ATAC-seq allows the classification of chromatin into open and closed regions, it is possible to achieve a much more fine-grained classification of chromatin states by analyzing a set of modifications of the DNA molecule that do not change the sequence, called *epigenetic modifications*. These include the methylation of the DNA molecule and a series of modifications of the histone proteins. In this section, we will concentrate on the use of histone modifications for chromatin state classification, also because this analysis will provide us with a perfect example of the application of *Markov models* to sequence analysis, a method that can be applied to many different problems.

Histone modifications consist mainly in the methylation or acetylation of specific aminoacids of the histone proteins. Many of them are associated to specific DNA states, determining its regulatory activity. For example, acetylation of lysine 27 of the core histone protein H3 (H3K27ac) marks active regulatory regions, and in particular active enhancers (i.e., regulatory regions that are distal from the gene they regulate); trimethylation of lysine 4 of the same protein (H3K4me3) is typically associated to active promoters, that is, gene-proximal regulatory regions, while the same modification on lysine 9 (H3K9me3) marks heterochromatic, inactive DNA regions.

As antibodies that recognize these specific histone modifications are available, they can be located on the genome sequence by ChIP-seq, just as the binding of transcription factors. Thus by mapping, for example, all the regions marked by H3K4me3 (i.e., the corresponding ChIP-seq peaks), we can obtain a catalog of the promoters that are active in a certain cell type or tissue. Then we could limit the search for candidate TF binding sites to these regions, in which the binding site would in fact be accessible to the TF. Similarly, we could use H3K27ac as a marker of active enhancers.

However, since several histone marks give complementary and partially independent information on the state of the chromatin, a more ambitious program is to integrate them so as to obtain a *segmentation* of the genome into regions with similar characteristics. Such segmentation would be cell type dependent, since the chromatin state of a region, and the corresponding histone marks, do change among cell types, reflecting the regulatory regions that guide the expression of the transcriptomic profile defining each cell type.

Markovian models, including Markov chains and hidden Markov models, provide us with a mathematical framework for the integration of different signals to obtain such a segmentation, thus we will dedicate the next subsections to introducing them, starting with examples of applications much simpler than histone marks.

5.3.3 Markov Chains

> ℹ **Markov chain**
>
> A *Markov chain* describes a system evolving in discrete time steps. The system can be in any state of a *state space* at each instant t of time. The system evolves in time through *stochastic* rules, and the probability of being in a given state at time $t + 1$ depends only on the state at time t.

The evolution of the system is dictated by a *transition matrix*:

> **i** **Transition matrix**
>
> The *transition matrix* of a Markov chain is a square matrix with as many rows and columns as there are states. The element T_{ij}, or $T(i \rightarrow j)$, contains the probability that the system will be in state j at time $t+1$ if it is in state i at time t.

To be concrete, let us consider a (very) simple model of weather forecasting. In our model the weather can only be in one of two states, either "sun" or "rain". If today is sunny, there is an 80% probability that tomorrow will be sunny too, while if today it rains, it will also rain tomorrow with a probability of 50%. These rules translate into the following transition matrix:

	sun	rain
sun	0.8	0.2
rain	0.5	0.5

Note that the sum of each row is equal to 1. This is necessary for the model to make sense, since the sum of the row "sun" is the sum of the probabilities of sun and rain after a sunny day, and therefore must be 1, etc. This is a consistency rule that applies to all Markov chain transition matrices:

$$\sum_j T(i \rightarrow j) = 1 \quad \forall i$$

The concept of time needs not to be interpreted literally: A Markov chain can describe any *ordered sequence* of events. Therefore Markov chains can be applied, in particular, to biological sequences, such as those of nucleic acids and proteins. For example we can define a Markov chain describing a DNA sequence with the following transition matrix:

	A	C	G	T
A	0.25	0.25	0.25	0.25
C	0.25	0.25	0.25	0.25
G	0.25	0.25	0.25	0.25
T	0.25	0.25	0.25	0.25

This is a rather trivial Markov chain: the probability of having any letter following any letter is simply 1/4, so that this Markov chain produces a completely random sequence. This is a sequence produced by this Markov chain:

TCAGAGCTACAGCCAGCCCTTGTTTATTCCGTACTAGCG

Let us change the transition matrix:

	A	C	G	T
A	0.45	0.05	0.05	0.45
C	0.05	0.45	0.45	0.05
G	0.05	0.45	0.45	0.05
T	0.45	0.05	0.05	0.45

Now an A or a T has 90% probability of transitioning to another A or T and only 10% of transitioning to G or C, and similarly for transitions from G and C. Here's a sequence produced by this Markov chain:

AAAAATTATCGGGCCAAAATTATATTGGCGGCGCCCGGGG

The following is an extreme example which actually produces a deterministic output:

	A	C	G	T
A	0	1	0	0
C	0	0	1	0
G	0	0	0	1
T	1	0	0	0

ACGTACGTACGTACGTACGTACGTACGTACGTACGTACGT

5.3.4 Markov Chains and CpG Islands

The actual human genome sequence does not look as if it was produced by any of these Markov chains (certainly not the last one, but also the completely random one is not a satisfactory description). A Markov chain which reproduces the transition probabilities for *most* of the human genome has the following transition matrix[7]:

	A	C	G	T
A	0.3	0.2	0.28	0.22
C	0.32	0.28	0.08	0.32
G	0.25	0.25	0.3	0.2
T	0.18	0.24	0.29	0.29

(this can be computed simply by counting how many times in the genome sequence an A is followed by another A, a C, etc.). We will refer to the Markov chain as $T^{(g)}$ (for genome).

---

[7]The treatment of CpG islands with Markovian models is taken from Chapter 3 of [8].

Notice in particular that the $C \rightarrow G$ transition probability is very low (0.08). The reason is that the dinucleotide CpG (i.e., a C followed by a G with a *phosphodiester* bond between the two: the notation is meant to distinguish it from the CG Watson-Crick pairing) is subject to frequent mutations after methylation that change it into TpG. However, there are regions in the genome which, possibly due to some evolved mechanism of protection from methylation, show a much higher frequency of CpGs than the rest of the genome. These are called *CpG islands*, extend typically for a few hundred base pairs, and are described by the Markov chain $T^{(CpG)}$:

	A	C	G	T
A	0.18	0.27	0.43	0.12
C	0.17	0.37	0.27	0.19
G	0.16	0.34	0.37	0.13
T	0.08	0.36	0.38	0.18

In agreement with the definition, this Markov chain has a much greater $C \rightarrow G$ probability (0.27) than that in the rest of the genome (0.08), but there are large differences between the two chains in all transition probabilities.

CpG islands are functionally important, as they often coincide with regulatory regions, so that it is important to have a mathematical method to identify them in the genome. A simple way of doing so is to count the CpG dinucleotides, but by exploiting the global differences in dinucleotide composition, we can obtain a better classifier. Clearly, Markov chains are the natural mathematical tool. Here we will see how to use Markov chains to solve the following problem:

given a genomic sequence of the typical length of a CpG island, determine whether it is a CpG island or not

Later in the chapter, we will see that the apparently similar but actually more complex problem:

identify all the CpG islands in the human genome

requires a more sophisticated type of Markov chain, called a hidden Markov model.

5.3.5 Likelihood of a Markov Chain

Markov chains are defined as *generative models*, that is, models that we can use to generate sequences that look like the actual genome sequence (within a CpG island or not, depending on which Markov chain we use). However, in most cases, they are not used to *generate* sequences but to *classify* existing sequences. The procedure is very similar to the one we used to assign a score to a candidate TF binding site based on a PPM, and is based on the *likelihood* of a Markov model given a sequence.

Just as with PPMs, given a DNA sequence S, we can compute the likelihood of a Markov chain with respect to the sequence.

> ℹ **Likelihood of a Markov chain**
>
> Given a Markov chain represented by a transition matrix T and a sequence of states S, the *likelihood* of T given S is the probability that T will generate S.

We just have to count how many times each transition occurs in the sequence and multiply the probabilities of all such transitions[8]. For example, the sequence:

$$S = GCGCG$$

contains two $G \to C$ transitions and two $C \to G$ transitions; therefore, the likelihood of the $T^{(g)}$ Markov chain is:

$$L(T^{(g)}|S) = T^{(g)}(G \to C)^2 \cdot T^{(g)}(C \to G)^2 = 0.25^2 \cdot 0.08^2 = 4 \cdot 10^{-4}$$

This number is not particularly interesting *per se*, but let us compare it with the value obtained for the same sequence using the T^{CpG} Markov chain:

$$L(T^{(CpG)}|S) = T^{(CpG)}(G \to C)^2 \cdot T^{(CpG)}(C \to G)^2 = 0.34^2 \cdot 0.27^2 = 8.43 \cdot 10^{-3}$$

which is about 21 times larger than $L(T^{(g)}|S)$.

As in the case of PPMs, the *likelihood ratio* gives us a quantitative way of comparing the two models:

$$LR = \frac{L(T^{(CpG)}|S)}{L(T^{(g)})|S}$$

Also in this case, the two models in the likelihood ratio are not nested, and thus the likelihood ratio cannot be directly converted into a P-value. However, it can be used as a classifier: A sequence S will be classified as a CpG island if $LR > 1$ and as non-CpG island otherwise. For our example $S = GCGCG$ the likelihood ratio is ~ 21, so we classify it as CpG island[9]. In practice, as for PPMs, we never compute likelihoods but always their logarithms because the likelihoods for long sequences are extremely small numbers, which create computational problems.

This procedure offers a solution to our first problem: given a sequence, determine whether it is a CpG island or not. In the following sections we will tackle the second problem, that is of identifying all CpG islands in a long sequence, such as a chromosome or the whole genome. A simple solution to the problem would be:

---

[8]To be precise, we would also have to specify the probability distribution of the first base. For simplicity, we will ignore this problem, noting that it becomes progressively less relevant as longer sequences are considered.

[9]As we could have easily guessed simply from the fact that it contains two CpGs.

- divide the genome into windows of fixed size (say 100 bp),
- use Markov chains to classify each window as CpG-island or non-CpG island, and
- merge consecutive windows of the same class.

While acceptable, this procedure suffers from some arbitrariness: Different choices of window size would lead to different results, identifying CpG islands of different size. Instead, we would like the size of each individual CpG island to emerge naturally from the procedure. Hidden Markov models offer a solution to this problem and, as we shall see below, to other problems of classification of genomic regions, including that of segmenting the genome based on histone marks.

5.3.6 Hidden Markov Models

In the examples of Markov chains that we have seen, the DNA sequence we observe coincides with the sequence of the states of the Markov chain.

> **i   Hidden Markov model**
>
> A *hidden Markov model* (HMM) is a Markov chain whose states are not directly observable: Instead, we observe *symbols* that are *emitted* by the states and give us some information about the underlying states.

What we can observe is a sequence of symbols, but what we are really interested in is the underlying, unsobservable sequence of states. The problem is thus to infer the latter from the former. Let us return to the weather example[10]. Our Markov chain is the same as before, with the states "sun" and "rain" and the transition matrix:

	sun	rain
sun	0.8	0.2
rain	0.5	0.5

However, suppose you are locked in a windowless room and have no way to look outside and see the weather. Every day, your friend tells you over the phone what she did that day[11]: walk, shop, or clean her house (every day she does exactly one of these activities). From the sequence of her activities (the *symbols*), you want to guess the weather (the *states*). This is possible if you know not only the transition matrix, but also the *emission probabilities*:

---

[10]This example is modified from the Wikipedia page "Hidden Markov model."

[11]You *do not* want your friend to know that you are locked in a room, so you cannot simply ask her about the weather.

> ℹ️ **Emission probabilities**
>
> The *emission probabilities* of a HMM specify the probability that each state will *emit* each symbol.

In our case, the emission probabilities specify the probability that your friend will engage in any of the three activities when the weather is sunny or rainy. Suppose these are the emission probabilities:

	walk	shop	clean
sun	0.6	0.3	0.1
rain	0.1	0.4	0.5

The sum of each row is again 1, indicating that she always performs exactly one of the three activities[12]. Clearly, given a sequence of symbols, you cannot infer the sequence of states with certainty. However, intuitively, if you observe the sequence of symbols

walk, walk, walk, clean, clean

then the sequence of states

sun, sun, sun, rain, rain

is much more likely than

rain, rain, rain, sun, sun

The process of guessing the sequence of states from the observed sequence of symbols is called *decoding*.

> ℹ️ **Decoding**
>
> Given a sequence s of symbols, *decoding* it means finding the sequence of states with the maximum likelihood given s.

The likelihood of a sequence of states given a sequence of symbols can be computed from the transition probability matrix (giving us the probability of the sequence of states), and the emission probability matrix (to compute the probability that the sequence of states will emit the observed symbols). For example, indicating with E the elements of the emission matrix, the likelihood of the sequence of states

---

[12]This is not a necessary requirement for a HMM, and indeed the HMM that we describe below for chromatin segmentation does not have this property.

sun, sun, sun, $rain$, $rain$

given the sequence of symbols

$walk$, $walk$, $walk$, $clean$, $clean$

is given by:

$$
\begin{aligned}
L(sun &- sun - sun - rain - rain | walk - walk - walk - clean - clean) \\
&= E(walk|sun)T(sun \to sun)E(walk|sun)T(sun \to sun) \\
&\quad \times\ E(walk|sun)T(sun \to rain)E(clean|rain) \cdot T(rain \to rain)E(clean|rain) \\
&= E(walk|sun)^3 E(clean|rain)^2 T(sun \to sun)^2 T(sun \to rain)T(rain \to rain) \\
&= 0.6^3 \cdot 0.5 \cdot 0.8^2 \cdot 0.2 \cdot 0.5 = 0.0069
\end{aligned}
$$

Thus, in principle one could perform the decoding by computing the likelihood of every possible sequence of states and choosing the one with the highest likelihood. Obviously, this is practically unfeasible except for very short sequences. Luckily there are algorithms, such as the Viterbi algorithm, based on dynamic programming (just like the exact alignment algorithms mentioned in Chapter 1), which produce the exact solution in a short time even for long sequences.

5.3.7 A Hidden Markov Model for CpG Islands

Returning to CpG islands, we define a HMM with:

- eight states $A_+, C_+, G_+, T_+, A_-, C_-, G_-, T_-$
- four symbols A, C, G, T

The emission probabilities are trivial: A_+ and A_- always emit the symbol A, etc. The transition probabilities are given by the following matrix:

	A_+	C_+	G_+	T_+	A_-	C_-	G_-	T_-
A_+	$0.18p$	$0.27p$	$0.43p$	$0.12p$	$\frac{1-p}{4}$	$\frac{1-p}{4}$	$\frac{1-p}{4}$	$\frac{1-p}{4}$
C_+	$0.17p$	$0.37p$	$0.27p$	$0.19p$	$\frac{1-p}{4}$	$\frac{1-p}{4}$	$\frac{1-p}{4}$	$\frac{1-p}{4}$
G_+	$0.16p$	$0.34p$	$0.37p$	$0.13p$	$\frac{1-p}{4}$	$\frac{1-p}{4}$	$\frac{1-p}{4}$	$\frac{1-p}{4}$
T_+	$0.08p$	$0.36p$	$0.38p$	$0.18p$	$\frac{1-p}{4}$	$\frac{1-p}{4}$	$\frac{1-p}{4}$	$\frac{1-p}{4}$
A_-	$\frac{1-q}{4}$	$\frac{1-q}{4}$	$\frac{1-q}{4}$	$\frac{1-q}{4}$	$0.30q$	$0.20q$	$0.28q$	$0.22q$
C_-	$\frac{1-q}{4}$	$\frac{1-q}{4}$	$\frac{1-q}{4}$	$\frac{1-q}{4}$	$0.32q$	$0.28q$	$0.08q$	$0.32q$
G_-	$\frac{1-q}{4}$	$\frac{1-q}{4}$	$\frac{1-q}{4}$	$\frac{1-q}{4}$	$0.25q$	$0.25q$	$0.30q$	$0.20q$
T_-	$\frac{1-q}{4}$	$\frac{1-q}{4}$	$\frac{1-q}{4}$	$\frac{1-q}{4}$	$0.18q$	$0.24q$	$0.29q$	$0.29q$

The interpretation is the following: A_+ represents an A inside a CpG island, A_- an A outside of a CpG island, and so on. Therefore, we can indeed observe the symbols (the DNA sequence) but we need to decode the sequence of symbols to learn the sequence of states, that is, which parts of the DNA sequence are CpG islands and which are not.

The transition probabilities among + states are just those of our Markov chain describing CpG islands, multiplied by a new parameter p. Similarly, those among − states are the same as our Markov chain $T^{(g)}$ multiplied by q. It follows that a + state will transition to another + state with probability p, and will choose which one according to the transition probabilities of CpG islands; it will transition to a − state with probability $1 - p$, and will choose the − state completely at random. Same for − states, with p replaced by q.

Therefore, our Markov chain will evolve just like a CpG island for a series of steps, then transition into a non-CpG island and behave like one for a while, then turn into a CpG island again, and so on, precisely how we imagine the real genome behaves. The typical length of a CpG island generated by this model will be $1/(1-p)$ (as on average it will transition into a non-CpG island after $1/(1-p)$ steps) and that of a typical stretch of regular genome will be $1/(1-q)$. Therefore, the parameter $1-p$ will be set to be the inverse of the typical size of a CpG island, and $1-q$ to the inverse of the typical distance between consecutive CpG islands.

Having established the model, we use it to classify the actual genome sequence through decoding, that is we find the sequence of states corresponding with maximum likelihood to our observed sequence of symbols. The problem is simplified (compared with the weather problem above) by the fact that the emission probabilities are deterministic, so, for example, the likelihood:

$$L(T_- A_- G_+ G_+ T_+ | CAGGT)$$

is trivially 0 since T_- cannot emit a C. Therefore, only the transition probabilities enter the likelihood calculations. For example:

$$L(C_- A_- G_+ G_+ T_+ | CAGGT) = T(C_- \rightarrow A_-) T(A_- \rightarrow G_+) T(G_+ \rightarrow G_+) T(G_+ \rightarrow T_+)$$
$$= 0.32q \cdot \frac{1-q}{4} \cdot 0.37p \cdot 0.13p$$

Decoding can be performed with the Viterbi algorithm. The regions where the state was decoded as A_+, C_+, G_+, T_+ are classified as CpG islands, and those decoded as A_-, C_-, G_-, T_- as non-CpG-islands. Just as we were inferring the weather from the activities of our friend, we infer the location of the CpG islands from the transition patterns between bases.

HMMs are used in many different scientific and technological fields, such as speech recognition. Here, roughly speaking, the symbols are the sounds recorded by a microphone (suitably transformed mathematically), and the states are the words that generated those sounds. We will have a matrix of transition probabilities from one word to the next word, and emission probabilities from words to sounds. Decoding will tell us which sequence of words is most likely to have generated the sequence of sounds that the microphone recorded. In the next subsection, we will describe an application of HMMs to classify the genome by the state of the chromatin.

5.3.8 ChromHMM: Segmenting the Genome into Chromatin States

Here, we describe an approach developed in [11] to classify the human genome by function, and in particular to locate the regulatory regions, based on histone marks measured by ChIP-seq. The basic idea is to use a HMM in which:

- the (hidden) states describe the possible regulatory states of the chromatin (e.g., we will have enhancers, promoters, and heterochromatin among our states);
- the transition matrix describes how these states follow each other on the genome;
- the symbols, that we can measure, are the ChIP-seq signals of a set of histone marks;
- the emission probabilities describe the probability with which each state produces each histone mark; and
- decoding will allow us to establish which particular sequence of states is most likely to have generated the combination of ChIP-seq signals that we observe.

5.3.9 The Data, the HMM, and the States

The data consist of nine chromatin marks (eight histone modifications and CTCF binding[13]) measured by ChIP-seq in nine human cell lines. In the CpG island example discussed above, the states of the HMM were established a priori, and the transition matrix was derived from pre-classified data. Here also the HMM states and their emission probabilities were obtained from the data through an optimization procedure, described in detail in a previous publication [10]. The authors of [11] use a 15-state HMM. Since the states are determined by the data, they must be interpreted, to be useful, in terms of meaningful categories such as "enhancer," "promoter," "insulator," etc. This can be done by inspecting the emission probabilities, which are shown in Figure 5.8. Note that, contrary to our previous examples of HMMs, the rows of this emission probability need not sum to 1, since each state can simultaneously emit multiple symbols, that is, be associated to multiple histone marks.

For example, state 1 emits the histone marks known to be associated to active promoters, and can thus be interpreted as such, while states 4 and 5 can be interpreted as representing enhancers. State 12 emits almost only H3K27me3, known to be associated to repression by the Polycomb complex, and is therefore named "Polycomb repressed."

The decoding process will provide a *segmentation* of the genome in which each base is assigned to one of the 15 states. Since epigenetic modifications are tissue- and cell-type-specific, so will be the genome segmentation. Figure 5.9 shows the genomic region shown before in Figure 5.4, with POU5F1 peaks near the NANOG gene, together with its ChromHMM segmentation in several cell lines, using the color

---

[13]As we will see later in this chapter, CTCF is a quite special TF which plays a fundamental role in the architecture of the chromatin.

class	CTCF	H3K27me3	H3K36me3	H4K20me1	H3K4me1	H3K4me2	H3K4me3	H3K27ac	H3K9ac	
	0.16	0.02	0.02	0.06	0.17	0.93	0.99	0.96	0.98	1 active promoter
	0.12	0.02	0.06	0.09	0.53	0.94	0.95	0.14	0.44	2 weak promoter
	0.13	0.72	0.00	0.09	0.48	0.78	0.49	0.01	0.10	3 inactive/poised promoter
	0.11	0.01	0.15	0.11	0.96	0.99	0.75	0.97	0.86	4 strong enhancer
	0.05	0.00	0.10	0.03	0.88	0.57	0.05	0.84	0.25	5 strong enhancer
	0.07	0.01	0.01	0.03	0.58	0.75	0.08	0.06	0.05	6 weak/poised enhancer
	0.02	0.01	0.02	0.01	0.56	0.03	0.00	0.06	0.02	7 weak/poised enhancer
	0.92	0.02	0.01	0.03	0.06	0.03	0.00	0.00	0.01	8 insulator
	0.05	0.00	0.43	0.43	0.37	0.11	0.02	0.09	0.04	9 transcriptional transition
	0.01	0.00	0.47	0.03	0.00	0.00	0.00	0.00	0.00	10 transcriptional elongation
	0.00	0.00	0.03	0.02	0.00	0.00	0.00	0.00	0.00	11 weak transcribed
	0.01	0.27	0.00	0.02	0.00	0.00	0.00	0.00	0.00	12 polycomb repressed
	0.00	0.00	0.00	0.00	0.00	0.00	0.00	0.00	0.00	13 heterochromatin/low signal
	0.22	0.28	0.19	0.41	0.06	0.05	0.26	0.05	0.13	14 repetitive/copy number variation
	0.85	0.85	0.91	0.88	0.76	0.77	0.91	0.73	0.85	15 repetitive/copy number variation

Scale: 0.8, 0.6, 0.4, 0.2, 0

Figure 5.8 Emission probabilities of the 15 ChromHMM states. The rows represent the 15 states and the columns the 9 symbols. Each cell in the matrix represents the probability that a state will generate the signal corresponding to the symbol. Note that each state typically produces multiple symbols, so the rows need not sum to one. States are further organized into classes (colors on the left side) for visualization in the genome browser (see below Figure 5.9).

coding established in Figure 5.8. Thus the POU5F1 peaks are located in a region whose chromatin is in an active regulatory state specifically in embryonic stem cells (precisely the cells in which the ChIP-seq assay for POU5F1 binding was carried out), while it is inactive in all other cell lines considered.

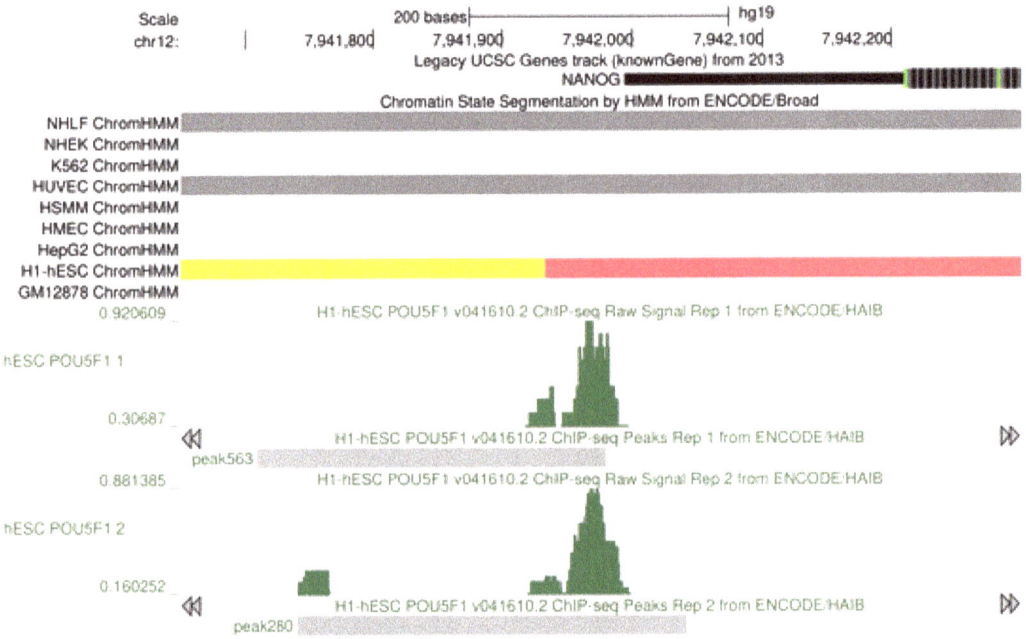

Figure 5.9 Segmentation of the genome into chromatin states using ChromHMM for the same region shown in Figure 5.4. This region is in inactive states (light and dark gray) in all cell lines except for human embryonic stem cells (H1-hESC), where the region is in active, regulatory states (red and yellow). Indeed the POU5F1 peaks were found from ChIP-seq in precisely this cell line.

5.4   THE THREE-DIMENSIONAL CONFORMATION OF THE GENOME

5.4.1   Chromosomal Contacts

Besides the chromatin states discussed before, another property of the genome relevant to gene regulation is the 3D conformation of the DNA molecules inside the cell nucleus. The complex packing of DNA makes it possible for regions that are far away in the linear sequence to come into close proximity, allowing physical interactions to occur between transcription factors bound to the two regions. This is the mechanism behind the action of distal regulatory regions, such as enhancers, on gene expression. Therefore, knowing which physical interactions are made possible by the 3D conformation of the genome helps understanding which genes are the regulatory targets of each enhancer. As we will see in Chapter 6, this is crucial, in particular, to understand mechanistically the genetic bases of complex traits and diseases.

Hi-C is a relatively recent addition to a series of technologies for chromosome conformation capture, and the first technique to allow such mapping at the scale of the whole genome. Like RNA-seq and ChIP-seq, it is another example of how next generation sequencing techniques revolutionized not only the study of the DNA sequence and its variation, but also that of the epigenetic components of gene regulation. In Hi-C, the genome is cross-linked to "freeze" the interactions between loci. After cutting with a restriction enzyme, the ends of the originally interacting fragments are ligated so that regions that were originally non-contiguous in the linear DNA sequence, but physically interacting, are now ligated into a single sequence. Next-generation sequencing is applied to the fragmented DNA: Specifically, paired-end sequencing is used so that a pair of reads aligned to two stretches of the genome that are not close together in the linear genome implies that the two stretches were physically interacting in the 3D genome.

To build a map of these interactions, the genome is first divided into *loci* of equal size, and a *contact map* is defined:

ℹ **Hi-C contact map**

The elements M_{ij} of the *contact map* M indicate the probability that loci i and j are in physical proximity in the 3D genome.

Figure 5.10 represents the contact matrix of a ~900 Kb region in chromosome 3, for a lymphoblastoid cell line, as shown in the UCSC genome browser.

A few features are immediately evident: First, the diagonal elements of M contain high contact values in all loci because loci that are close in the linear genome are necessarily also close in 3D space, and hence often in contact. Moreover, most of the off-diagonal signal is concentrated in squares along the diagonal, called *contact domains*, whose typical length is a few hundred Kb:

ℹ **Contact domain**

A *contact domain*, represented by a square of high values along the diagonal of the contact matrix, represents a region of the genome in which all loci interact with each other much more than they interact with loci outside the domain.

Finally, some contact domains, such as the largest one in Figure 5.10, show small regions of stronger enrichment at their corners, called *peaks*, which indicate strong contact between the two ends of the domain:

ℹ **Peaks and loops**

A *peak* is a localized enrichment of contacts which can be interpreted as a *loop* joining the two ends of a contact domain.

Thus, the two ends of these contact domains form a DNA loop allowing them to interact more strongly than the interior of the domain. Peaks/loops can also be found in loci other than those delimiting contact domains, and often put an enhancer in

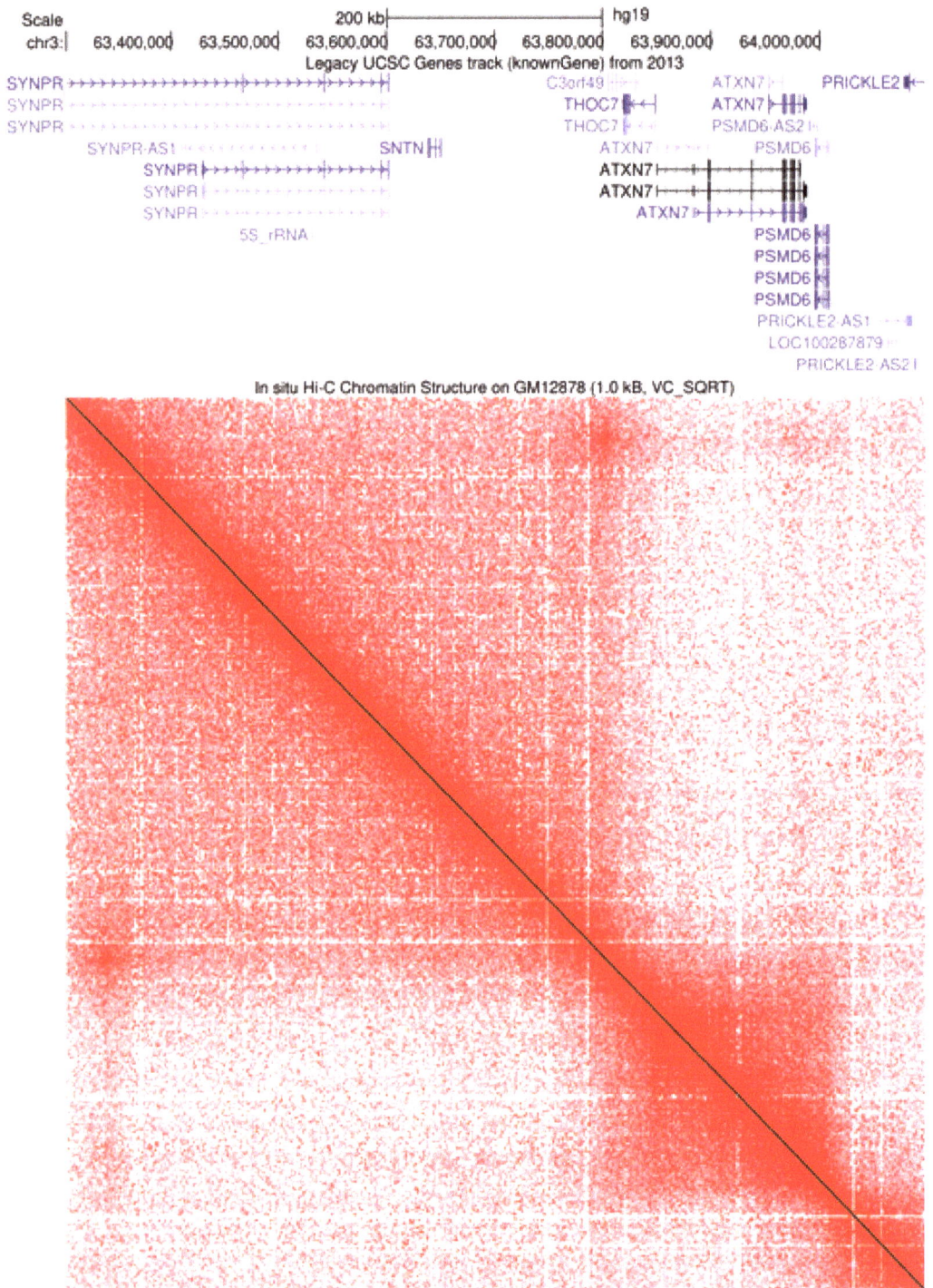

Figure 5.10 Heatmap representation of the contact map for a ∼900 Kbp region on chromosome 3, derived from a Hi-C assay on a lymphoblastoid cell line. The y axis in the heatmap represents (from top to bottom) the same region shown in the x axis, and the intensity of the red color is proportional to the strength of the interaction between each pair of loci (here the resolution, i.e. the size of each locus, is 10 Kb).

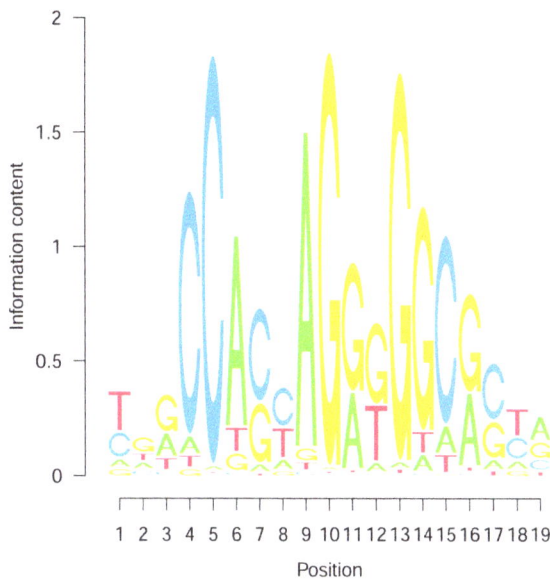

Figure 5.11 Logo of the PFM associated by the JASPAR database to CTCF.

contact with a promoter, thus enabling their physical interaction and the regulatory effect of the enhancer on gene expression.

5.4.2   The Role of CTCF

CTCF (CCCTC-binding factor) is a transcription factor highly conserved in bilaterian animals[14], where it is ubiquitously expressed in virtually all tissues. While its role in chromatin organization has been known for some time, a precise mechanism by which CTCF binding determines 3D genome conformation emerges from the combined analysis of ChIP-seq data on CTCF binding and Hi-C data.

First, CTCF binding sites tend to be overrepresented at the two ends of the chromatin loops identified by Hi-C. However, the precise mechanism by which CTCF plays a role in the formation and maintenance of such loops becomes clearer when we analyze the *relative orientation* of the CTCF binding sites found on the two ends of the loop.

Suppose that by combining ChIP-seq and Hi-C data we found a list of chromatin loops that are associated with CTCF binding sites, and we identified the CTCF motif at each end of the loop (using the PFM, shown in Figure 5.11, representing the sequence specificity of CTCF). For simplicity, let us consider only positions 4, 5, 9, and 10 in the CTCF PFM of Figure 5.11 and represent the CTCF motif by CC...AG... . Since the motif can be on either strand, it can actually be represented by CC...AG... or by its reverse complement CT...GG..., so that the DNA sequence around the chromatin loop can take four possible forms:

---

[14]Although it has been lost in some lineages, in particular in the model nematode *Caenorhabditis elegans*.

```
...CC...AG...(loop)...CC...AG...
...CC...AG...(loop)...CT...GG...
...CT...GG...(loop)...CC...AG...
...CT...GG...(loop)...CT...GG...
```

which correspond to four possible orientations of the bound CTCF molecules with respect to each other and the loop. Let us represent the CTCF molecule as an arrow pointing from the region binding "CC" to the region binding "AG". Detailed biochemical studies of the binding of CTCF to DNA show that this corresponds to an arrow going from the C- toward the N-terminus of the protein. The four DNA sequences represent four orientations of the two CTCF molecules:

```
      =>                      =>
...CC...AG...(loop)...CC...AG...
      =>                      <=
...CC...AG...(loop)...CT...GG...
      <=                      =>
...CT...GG...(loop)...CC...AG...
      <=                      <=
...CT...GG...(loop)...CT...GG...
```

If the orientation of the CTCF molecule were irrelevant to its role in chromatin loops, the four situations would be equally probable. However, analyzing the data reveals that the second possibility (*convergent* CTCF binding sites) is by far the most common, being realized in ~90% of the cases. This suggests that the effect of CTCF binding on the formation of chromatin loops depends on the relative orientation of the two binding sites at the two ends of the loop: A convergent orientation of the two CTCF sites seems required to foster the creation of the loop[15]. Further studies have interpreted this orientation requirement as allowing the formation of the loop through a process called "loop extrusion" [34].

To summarize, both chromatin states derived from histone marks and the 3D conformation of the genome derived from Hi-C data help us understand which regions are able to regulate which genes. In Chapter 6, we will see how these tools become critical in the study of complex traits and diseases, as it turns out that the genetic variation underlying these phenotypes is mostly located in regulatory regions of the genome.

FURTHER READING

Durbin, R., Eddy, S., Krogh, A. & Mitchison, G. Biological Sequence Analysis. (Cambridge University Press,1998) - Chapter 3.

---

[15] We call this configuration "convergent" because we *arbitrarily* assigned the CTCF molecule orientation as described. With the opposite convention we would have said that divergent sites are needed at the two ends of chromatin loops. In both cases we can conclude that the N-termini of both CTCF proteins must face toward the chromatin loop.

Ernst, J., Kheradpour, P., Mikkelsen, T., Shoresh, N., Ward, L., Epstein, C., Zhang, X., Wang, L., Issner, R., Coyne, M., Ku, M., Durham, T., Kellis, M. & Bernstein, B. Mapping and analysis of chromatin state dynamics in nine human cell types. *Nature.* **473**, 43–49 (2011).

Rao, S., Huntley, M., Durand, N., Stamenova, E., Bochkov, I., Robinson, J., Sanborn, A., Machol, I., Omer, A., Lander, E. & Aiden, E. A 3D map of the human genome at kilobase resolution reveals principles of chromatin looping. *Cell.* **159**, 1665–1680 (2014).

Sanborn, A., Rao, S., Huang, S., Durand, N., Huntley, M., Jewett, A., Bochkov, I., Chinnappan, D., Cutkosky, A., Li, J., Geeting, K., Gnirke, A., Melnikov, A., McKenna, D., Stamenova, E., Lander, E. & Aiden, E. Chromatin extrusion explains key features of loop and domain formation in wild-type and engineered genomes. *Proceedings Of The Nat Acad Sci.* **112**, E6456–E6465 (2015).

Schmidt, D., Schwalie, P., Wilson, M., Ballester, B., Gonçalves, A., Kutter, C., Brown, G., Marshall, A., Flicek, P. & Odom, D. Waves of retrotransposon expansion remodel genome organization and CTCF binding in multiple mammalian lineages. *Cell.* **148**, 335–348 (2012).

Zhang, Y., Liu, T., Meyer, C., Eeckhoute, J., Johnson, D., Bernstein, B., Nusbaum, C., Myers, R., Brown, M., Li, W. & Liu, X. Model-based analysis of ChIP-Seq (MACS). *Genome Biol.* **9**, R137 (2008).

CHAPTER 6

Variation and Phenotype

6.1 INTRODUCTION

In the previous chapters, we have examined several applications of NGS to the analysis of gene regulation (through techniques such as ChIP-seq and Hi-C) and its end product, namely gene expression (through RNA-seq). In this lecture, we discuss the analysis of the DNA sequence itself: The huge decrease of the cost of sequencing brought about by NGS has allowed the sequencing of very large samples of individuals and thus the possibility of studying systematically and quantitatively *genetic variation*, defined as the set of differences in DNA sequence between individuals of the same species, and its phenotypic effects.

Phenotypes (including diseases) with a genetic component can be divided into two classes: Mendelian and complex. These two classes are associated to very different patterns of genetic variation, and the computational methods used to study them are distinct. Therefore, we will discuss them separately. However, the main question we are interested in answering is the same: Given a phenotype with a genetic component, we want to identify (a) the genetic variants that cause it and (b) the molecular mechanisms by which the causal variants affect the phenotype.

6.2 GENETIC VARIANTS

6.2.1 Variant Classification

> **i Genetic variant**
>
> A genetic *variant* is any difference in the DNA sequence between individuals of a population.

Specifically, we are interested here in *germline variants*, which originate from mutations in the germ cells and can thus be transmitted to the descendants of the individual in which they first occurred. In this chapter, we will be concerned with those variants that affect the phenotype, and in particular those that cause or predispose to a disease. Genetic variants can be classified in at least three ways, all of which are relevant for our main concern:

DOI: 10.1201/9781003449928-6

1. Variants can be classified based on how the DNA sequence differs among individuals: In *single nucleotide variants* (SNVs), the difference is limited to one nucleotide in a specific position; in *insertion/deletion* variants, one or more nucleotides at a given position are present in some individuals and absent in others; a *copy number variant* indicates a genomic region present in a different number of copies in different individuals; finally, in *translocations and inversions* a genomic region is found in different position and/or orientation in different individuals.

2. Variants can be classified based on their *frequency* in the population, indicated by the *minor allele frequency* (MAF), that is, the frequency in the population of the less common version of the sequence. Since we now have DNA sequencing data for millions of individuals, the MAF of known variants vary between $\sim 10^{-6}$ and 0.5. Often a variant is defined to be *common* if its MAF is >1%. Common variants are often called *polymorphisms*, and in particular common SNVs are known as *single nucleotide polymorphisms* (SNPs)[1].

3. Finally, variants can be classified based on their location with respect to genes and exons, and their effect on protein sequence:

- *exonic* variants can be *coding* if they reside within the coding sequence; in this case they can be further classified as *non-synonymous* or *synonymous* depending on whether they change or not the protein sequence; *non-synonymous* variants can be *missense* (change a single aminoacid) or *nonsense* (introduce a stop codon which prematurely terminates translation)

- *intronic* variants

- *intergenic* variants

We will need a few more definitions: Most variants are *biallelic*, meaning that only two of the possible four nucleotides (*alleles*) are present in the population. The allele shown by the officially released human genome sequence is called the *reference allele*, and the other the *alternate allele*. The reference allele is often, but by no means always, the more common (*major*) one. Remembering that each individual carries two copies of the genome (at least for the autosomal chromosomes), the *genotype* of an individual at a variant can be defined as the number of copies of the alternate allele carried by the individual, which can take the values 0, 1, or 2.

6.3 FINDING CAUSAL VARIANTS IN MENDELIAN DISEASES

Diseases caused by mutations in a single gene are called Mendelian, since when they run in families, their inheritance is governed by Mendel's laws[2], implying that the

---

[1] A potentially confusing fact is that the main database of SNVs is called "dbSNP," but most of the variants it contains are actually rare. It is important to keep in mind that the fact that a variant is contained in dbSNP *does not* imply it is common.

[2] A disease does not need to run in families to be of genetic origin. *De novo* mutations that appear in the parents' germline can cause genetic diseases in the offspring without the disease having ever

genetic cause is a single mutation. Identifying the gene whose mutation is responsible for a Mendelian disease is obviously of great importance, in particular, because it can be included in screening tests and suggest therapeutic strategies. From many Mendelian diseases for which we know the causal mutation we learned that these mutations, with a few exceptions, are located in the coding part of the genome and are non-synonymous. Therefore *exome sequencing* is the best tool for this investigation: In this technique, DNA sequencing is preceded by *target enrichment* in which the genomic DNA is enriched for regions matching a predefined set of sequences, typically all the known exons, in which case we are performing *whole exome sequencing* (WES). In other cases the enrichment is further restricted to a subset of the known exons, for example, those of a *panel* of genes previously known to be involved in the disease of interest.

Suppose then that we have performed WES on a patient of a disease that we suspect to be of Mendelian genetic origin (either because of its pattern of inheritance, or because it shows some typical features of Mendelian diseases, such as developmental disruptions and early onset). The first, natural thing to do is to select all the non-synonymous variants. Moreover, we can limit the analysis to rare variants: Since each Mendelian disease is generally rare in the population[3], we do not expect a common variant to be causal.

While these criteria greatly restrict the search, it turns out that a typical human being, not affected by any Mendelian disease, carries hundreds of rare, non-synonymous variants. If we know (e.g., from family history) that the disease is recessive, we can limit the search to homozygous variants, which greatly reduces the candidates (although $5 \sim 10$ rare, non-synonymous, homozygous variants are present in a typical healthy subject). If we cannot assume that the disease is recessive, we are faced with the daunting task of finding the causal mutation among hundreds of candidates.

6.3.1 Gene Prioritization

Most computational tools that have been developed to help in this task use the concept of *gene prioritization*: By using gene features known to be associated to genes involved in Mendelian diseases in general, and especially in those similar to the disease of interest, we produce a ranked list of the genes found mutated in the patient, in which genes that are the most likely to be causal[4] of the disease are ranked on top. While this does not provide a complete solution of the problem, it generates ranked testable hypotheses for validation (e.g., in cell lines, animal models, or organoids), thus allowing to prioritize the experimental work.

Many tools to perform gene prioritization have been developed, but most share a basic architecture that operates on a list of *candidate genes* (the ones found mutated

---

been observed in the family. However, the individual carrying the mutation can transmit it to its own offspring, so that the disease will follow Mendel's laws in future generations.

[3]Although since there are thousands of such diseases, the health burden they pose is very significant: The total number of patients ranges in the hundreds of millions and the symptoms are often very severe.

[4]For brevity we call "causal gene" the gene containing the mutation that is causal of the disease.

in the patient), a list of *seed genes* (previously known to be involved in the disease at hand, or in similar diseases, or in related biological processes, etc.), and a collection of *data sources* believed to be useful in the identification of the causal gene. Data sources are typically of two forms:

(a) gene annotation databases (e.g.: Gene Ontology or other functional annotation databases; protein domains; target genes of known drugs and compounds)

(b) databases of gene-gene relationship (e.g.: protein-protein interactions; coexpression; co-occurrence in literature abstracts)

6.3.2   An Example: Endeavour

We will describe in some more detail one of the popular prioritization tools, Endeavour [38][5]. The user of this tool is required to provide a list of seed genes and a list of candidate genes. Endeavour uses 43 data sources to build as many *submodels* of the disease. For example, one of the data sources is the Gene Ontology annotation of human genes. The corresponding submodel is based on the seed genes provided by the user: With the exact Fisher test described in Chapter 3, the Gene Ontology terms enriched in the seed genes compared to the rest of the genome are identified. This list of enriched terms, together with their enrichment P values, constitute the Gene Ontology submodel of the disease. Then the candidate genes are scored: To each candidate gene a score is assigned depending on the terms to which it is annotated that are included in the submodel, and their enrichment P values. Thus, the candidate genes are ranked based on whether they are annotated to biological processes that we can assume to be relevant to the disease, since they are enriched in the seed genes. Similar procedures, adapted to the type of information encoded by each of the 43 submodels, allow to assign a score to each candidate gene for each submodel.

Therefore, if all data sources are used (the user can choose to use only a subset of those available), this step returns 43 scorings of the candidate genes, that need to be combined in a single, final score. This is done by first transforming each scoring into a ranking, and then computing a statistic based on the combination of the 43 rankings. A P-value can be associated to this statistic (the details of how this is done are somewhat beyond the mathematical level of these lectures). Finally, the candidate genes are ranked by this P-value.

An example of usage of Endeavour is shown in Figure 6.1: The tool was used in [9] to identify the gene causative of hereditary spastic paraparesis in a single family. Exome sequencing of an affected family member and his parents was followed by a complex process in which most variants were eliminated as unlikely to be causal based on the recessive pattern of inheritance of the disease, the frequency of the variants in the population, and other considerations. After this process, the authors

---

[5]Endeavour is not limited to the problem of finding causal genes of Mendelian diseases, but more generally allows to rank a list of candidate genes to find those more likely to be involved in a given biological process, a Mendelian disease being just an example. Moreover, it can be used to rank candidate genes in six different species. In our discussion, we will assume we are using it to rank candidate genes in a human Mendelian disease.

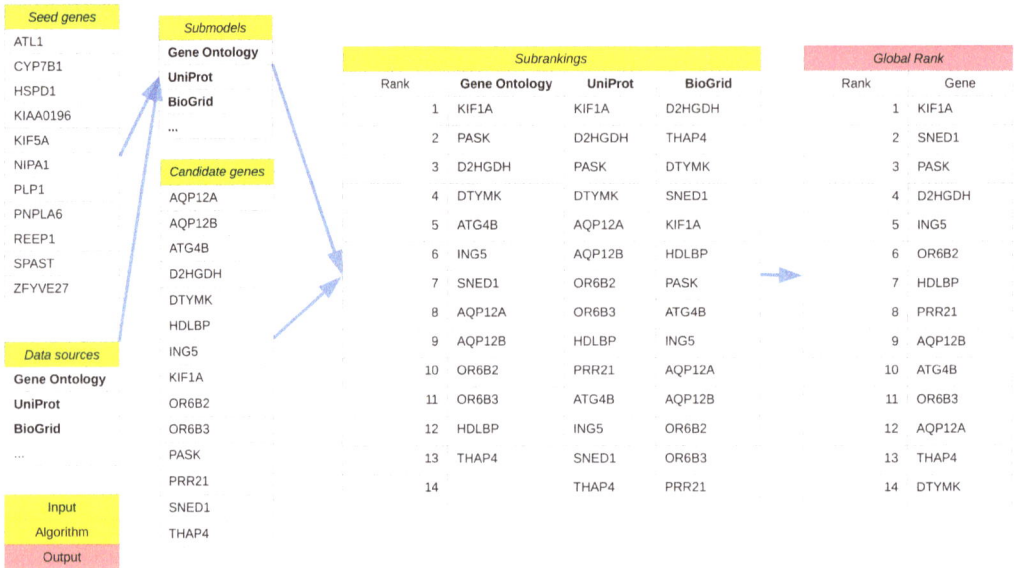

Figure 6.1 Use of Endeavour to identify the gene causal of hereditary spastic paraparesis in a family. Exome sequencing of one affected family member and his parents followed by genetic analysis led to a list of 14 candidate genes. Submodels for all the Endeavour data sources were generated based on a manually curated list of 11 seed genes, and used to produce subrankings of the candidate genes (not all candidate genes could be ranked in all submodels. Only three subrankings are shown). Finally the subrankings were combined into a single final ranking, in which KIF1A resulted as the most likely causal gene.

were left with 15 candidate variants, 14 of which were associated to a known gene; these 14 genes were fed to Endeavour as candidate genes. A list of 11 seed genes was manually curated from the literature. Endeavour identified *KIF1A* as the most likely to be causal among the candidate genes, and Sanger sequencing of the same gene in several family members confirmed that the variant was present in homozygosis in all and only the affected members, thus lending strong credibility to the hypothesis of causality.

6.3.3 Evaluating Performance: Cross-Validation

The process used by Endeavour, and similar processes used by other gene prioritization tools, transforms into a quantitative ranking the basic idea that the causal gene will be similar (across many definitions of similarity based on many data sources) to genes known to be involved in similar diseases or processes (the seed genes). This is biologically reasonable, but before using the tool to find new causal genes, we need to evaluate its performance. This can be done using *previously known* causal genes through *leave-one-out cross-validation*[6]. The basic idea is simple: Consider a disease

---

[6]Cross-validation is discussed more generally in subsection 6.4.6 below.

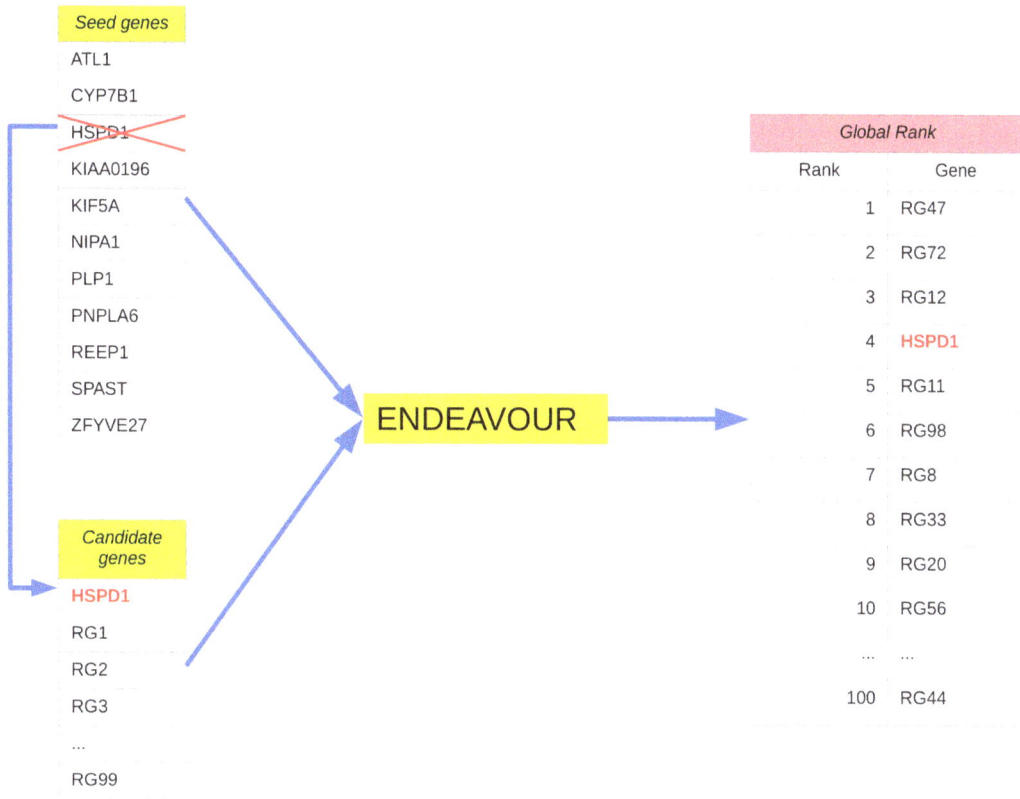

Figure 6.2 Leave-one-out cross-validation to evaluate the performance of Endeavour. One of the seed genes (*HSPD1*) is moved to the candidate gene list which contains 99 randomly chosen genes (*RG1 ... RG99*). Endeavour is run on these seed and candidate genes, and the final ranking of *HSPD1* is recorded. Intuitively, since *HSPD1* is a known causal genes and at least most of the random genes are not, Endeavour should place *HSPD1* near the top of the global ranking. This is repeated for all seed genes and their rankings are used to construct an ROC curve and compute the AUC.

and the genes that are already known to be causal for the disease, that is, the genes we would use as seed genes when looking for new ones. For each of these genes, *pretend* we do not know that it is causal, and thus remove it from the seed genes and place it among the candidate genes: How often the gene appears in the top-ranking ones provides a measure of the performance of the tool.

In the case of Endeavour, this is done as follows (see Figure 6.2):

(1) The seed genes for a Mendelian disease are extracted from the OMIM database, the main resource for known causal mutations in Mendelian diseases.

(2) Each seed gene in turn is removed and added to 99 randomly selected genes to create a list of 100 candidate genes.

(3) Endeavour is run (excluding OMIM as a data source) and the ranking of the removed seed gene is recorded.

Thus we end up with the rank of each seed gene, from which we can build an ROC curve (see Chapter 5) and compute the corresponding AUC as a measure of performance. The mean AUC of the Mendelian diseases used to validate the performance of Endeavour was 0.93, indicating very good performance (remember from Chapter 5 that a prefect classifier has AUC = 1, while a random classifier has AUC = 0.5).

6.4 COMPLEX TRAITS AND THE INTERPRETATION OF NON-CODING VARIANTS

6.4.1 Genome-Wide Association Studies

Many phenotypic traits, including diseases, are determined by both genetic and environmental factors. The genetic architecture of these traits is significantly different from that of Mendelian diseases: Rather than being driven by a single variant, they are typically determined by the cumulative effect of many variants whose individual effects are rather small. Due to such genetic architecture, these traits are known as *complex traits*, and include, for example, among phenotypic traits, height, body mass index, blood pressure, cholesterol level, etc.; and among diseases, diabetes, Crohn disease, schizophrenia, and many others that are in general much more common in the population than Mendelian diseases.

The genetic architecture of complex traits is investigated by *genome-wide association studies* (GWASs), in which the phenotype of interest is measured in a large cohort of individuals whose genotype is known. The genotype of the individuals can be assessed by whole genome sequencing (WGS) or, more often, using specific microarrays able to assess the genotype of an individual in millions of known polymorphic loci. The advantage of this technology with respect to WGS is its lower cost, an important factor given that GWASs typically require cohorts of many thousands of individuals. The disadvantage is that only known polymorphic loci can be assessed: However, since complex traits are mostly determined by common variants, this limitation is not critical.

> **i** **Genome-wide association study**
>
> A *genome-wide association study* (GWAS) identifies the genetic variants that are associated with a complex trait, that is, for which the number of copies of the variant carried by an individual is significantly correlated with the value of the trait. The trait can be binary (such as the presence of a disease) or continuous (any measurable phenotype expressed by a real number).

Therefore, in a GWAS, a statistical test is carried out for each measured variant to assess the significance of its correlation with the trait, using regression methods similar to those described in Chapter 2. For a *continuous* trait, expressed by a real number (height, blood pressure, body mass index, etc.), the association with each variant is assessed by linear regression, in which the dependent variable is the trait and the independent variable is the genotype (0, 1, or 2: number of copies of the alternate allele) of each individual at the variant of interest. Therefore, if y is the

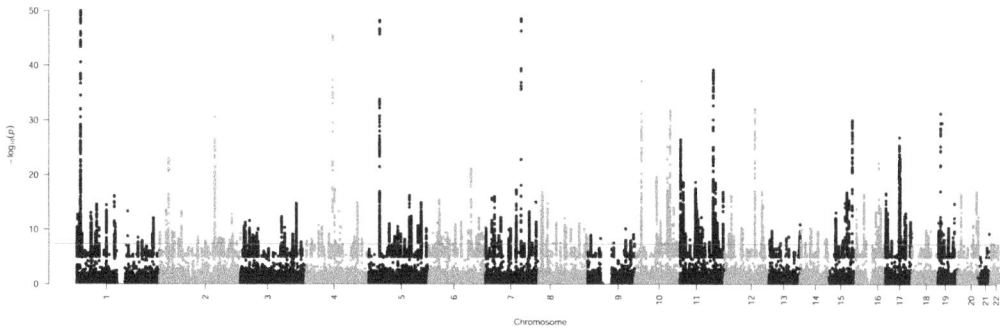

Figure 6.3 Manhattan plot of a GWAS performed on ~340,000 individuals, where the trait of interest is systolic blood pressure. The red line corresponds to $P = 5 \cdot 10^{-8}$, the significance threshold after multiple testing has been taken into account.

value of the trait and g is the genotype at a variant we assume[7]:

$$y = \beta_0 + \beta_g g + \epsilon$$

where the noise term ϵ represents the effect of all variables that affect the trait besides the variant of interest. The relevant quantity is β_g, which represents the change in the value of y due to each additional copy of the alternate allele compared with an individual homozygous for the reference allele. The P-value associated to β_g tests the null hypothesis $\beta_g = 0$ of no effect of the variant on the trait. Note that this approach assumes that the contribution of the maternal and paternal alleles to the trait are additive, that is, the absence of *dominance* effects.

This regression procedure is applied to all measured variants, and the P values are used to identify those that are significantly associated to the trait. Clearly, there is a massive multiple testing problem: For example, using a genotyping microarray capable of assessing 10^6 variants and Bonferroni correction, a variant would need a nominal association P-value of $0.05 \cdot 10^{-6} = 5 \cdot 10^{-8}$ to achieve an adjusted P-value of 0.05. The results of the association analysis are often displayed in a *Manhattan plot*[8], where the negative $\log_{10}$ of the P-value is displayed on the y axis while the x axis represents the linear genome, as in Figure 6.3.

Many traits of interest, and in particular complex diseases, are represented by a binary variable, with the values 0 and 1 representing, respectively, absence and presence of the disease. In this case linear regression cannot be used[9] and is replaced by *logistic regression*, which, like linear regression, associates to each variant a β_g coefficient and a P-value testing the null hypothesis $\beta_g = 0$. The coefficient β_g has the following interpretation: Each additional copy of the alternate allele multiplies

---

[7]This formula captures the fundamental idea but it is simplified with respect to actual GWAS studies, where multiple regression (see Chapter 2) is used to include, as covariates, other variables that could affect the trait y, such as age and sex, and possible *confounders*, namely, variables correlated to both the trait and the genotype, such as genetic ancestry.

[8]The name refers to the resemblance of these plots to the skyline of a modern city.

[9]Recall from Chapter 2 that linear regression assumes that the error term is normally distributed, which clearly cannot be the case for a binary variable.

the odds ratio[10] of the presence of the disease by a factor $\exp(\beta_g)$. Thus, if $\beta_g > 0$, the alternate allele is a *risk allele* (increases the odds of the disease), while if $\beta_g < 0$, the alternate allele is *protective*.

GWASs have been performed for thousands of different traits and found many significant associations between polymorphisms and traits, including diseases. As of March 20, 2024, the GWAS catalog repository contains more than 580,000 variant/-trait associations from almost 7,000 publications. For example, 25,548 associations have been found for the trait "body height" and 6,572 for "diabetes mellitus."

These associations can serve two related but distinct purposes:

(a) *Predicting* the genetic susceptibility of a trait or, in particular, of a disease. For example, if we can predict to what degree an individual is genetically predisposed to developing diabetes, we could target screening and prevention interventions in a more effective way. *Polygenic risk scores* summarize the information derived from a GWAS to assign a single number to each individual, based on their genotype in the GWAS-associated variants, which measures the genetic susceptibility to the disease.

(b) *Understanding* the mechanisms that drive genetic susceptibility might help in the development of therapies. For example, if many variants significantly associated to a disease are located near a gene, the gene might be involved in the pathogenic mechanism, and the corresponding protein might be a possible target of pharmaceutical intervention.

The two are quite independent, since we can have the ability to statistically predict the susceptibility of an individual to a disease without any mechanistic understanding of how the variants induce susceptibility. The task of understanding the mechanisms of genetic susceptibility is made particularly difficult by three facts:

(1) GWASs have confirmed that many variants are involved in each complex trait/disease, each individual variant having typically a very small effect. Obviously this makes it difficult to derive mechanistic hypotheses.

(2) The associated variants are mostly found in the non-coding part of the genome, and thus do not alter the sequence or structure of any protein. The most natural hypothesis is that these variants alter the phenotype by altering gene regulation and thus gene expression. The simplest case would be a variant in a non-coding, regulatory region, which alters the binding site of a TF and hence the regulation of one of its targets. As it should be clear from our discussion of transcription factor binding sites in Chapter 5, it is not easy to predict the effect of a variant in a regulatory region, since we do not understand the regulatory code as well as the genetic code. Moreover, since regulatory regions can be located very far

---

[10]If the probability of an event is p, its *odds* are defined as $\frac{p}{1-p}$. The *odds ratio* is the ratio of the odds for two genotypes. Thus the odds ratio of genotype 1 compared with genotype 0 is $\exp(\beta_g)$, which is also that of genotype 2 compared with genotype 1, while the odds ratio of genotype 2 compared with genotype 0 is $\exp(2\beta_g)$.

from the target gene (typically up to 1 Mb), it is not even clear which gene, if any, would have its expression affected by a variant.

(3) Finally, the phenomenon of *linkage disequilibrium* (LD) makes it difficult to understand which variants are actually *causally* related to the trait among those found *statistically associated* with the trait. The limited rate of genetic recombination leads to large segments of the genome (*LD blocks*) being passed to the later generations as a whole. This produces large correlations between the alleles of the variants within an LD block, so that if a variant in an LD block is causal for a trait, all the other variants in the block will be found statistically associated to the trait although they have no causal role. This is why Manhattan plots such as the one in Figure 6.3 show the characteristic "towers" of significant variants close to each other.

Fine mapping techniques have been developed to ameliorate problems (2) and (3). These techniques use both statistical tools and biological information to identify the variants most likely to play a causal role in the phenotype. Here, we will discuss how the study of transcriptomic data from large population samples can help fine mapping through the identification of *expression quantitative trait loci* (eQTLs).

6.4.2 eQTLs

> ℹ **Expression quantitative trait locus**
>
> An expression quantitative trait locus (eQTL) is a genetic variant statistically associated with the level of expression of a gene.

The identification of eQTLs is performed starting from a sample of individuals for which we know both the genotype (from WGS or microarrays) and the transcriptome (from gene expression microarrays or RNA-seq). Given a gene and a single-nucleotide variant[11], the analysis is performed in the same way as for any continuous trait, that is, by linear regression with gene expression as the dependent variable and the genotype at the variant as the independent one[12]. Therefore, we assume that the functional relationship between the expression E of the gene and the genotype g is:

$$E = \beta_0 + \beta_g g + \epsilon$$

where g is the genotype, and β_g represents the change in expression due to one more copy of the alternate allele. The P-value associated to each variant tests the null hypothesis $\beta_g = 0$.

While, in principle, a variant located anywhere in the genome can influence the expression of a gene (think, e.g., of a variant altering a transcription factor targeting

---

[11]We will consider, for simplicity, single nucleotide biallelic variants, but the principles apply to any type of variant, including multiallelic variants, insertions, deletions, etc..

[12]As in the case of GWAS, also here, in practice, multiple regression is used adding as covariates other variables that could affect gene expression, and possible confounders.

the gene of interest), most eQTL studies are limited to *cis-eQTLs*, that is, they analyze only the variants within a certain distance in the linear genome from the gene of interest (often of the order of 1 Mb). This choice is mainly motivated by the need to reduce the computational and multiple testing burdens by concentrating on the region where relevant variants are most likely to be located. Note that a truly comprehensive eQTL study, even if limited to the variants represented in the typical microarray and to protein-coding genes would entail $\sim 10^6$ variants $\times\, 2 \cdot 10^4$ genes $= 2 \cdot 10^{10}$ regression analyses.

As for GWAS, the assumption of linear dependence implies that the contribution of the paternal and maternal alleles to E are additive and independent. Although more sophisticated models have been developed to take into account possible dominance effects, most eQTL studies are performed in this framework, and there is evidence that dominance effects on gene expression are not widespread.

We will show an example from the GEUVADIS project [22], in which RNA-seq analysis was performed on lymphoblastoid cell lines from ~ 500 subjects whose genome had previously been sequenced within the 1000 Genomes project. Figure 6.4 shows the expression of gene *CSTB* as a function of the genotype of variant rs35285321 in 373 subjects of European ancestry, together with the best regression line.

The figure shows that the alternate allele A, which is less common than the reference allele G, leads to lower expression of CSTB, and the effect is greater for homozygous individuals. The very small P-value guarantees that the effect we see cannot be due to chance, thus this variant is a *bona fide* eQTL of CSTB. Of course this is a statistical association, and the problem of identifying causal variants is exactly the same as in GWASs.

6.4.3 An Example of Fine Mapping

We will discuss in some detail an example, taken from [12], which illustrates how the knowledge of chromatin states and eQTLs can guide the interpretation of GWAS results. SNP rs17293632 has been associated by large-scale GWASs with a series of inflammatory diseases including Crohn's disease. Moreover, *genetic fine mapping* analysis, a set of statistical methods, based on multiple regression, used to identify the most likely causal SNPs among those associated to a given trait, suggests this is indeed a good candidate to be causal of the disease.

We can learn something more through *epigenetic fine mapping*, that is, by using epigenetic data to understand the possible effects of this variant. The genomic and epigenomic context of this SNP is shown as a genome browser screenshot in Figure 6.5. The SNP is in an intron of the *SMAD3* gene, and is thus a non-coding variant not expected to change the sequence of any protein. However, the chromHMM tracks (Chapter 5) show that in all cell lines the analyzed SNP is within an active regulatory region, classified by chromHMM as either an enhancer (yellow) or a promoter (red). Thus, we can hypothesize that the effect of the variant is to change the regulation and hence the expression level of a gene, possibly *SMAD3*, which in turn can affect the phenotype, and in particular susceptibility to the disease. This could happen, for

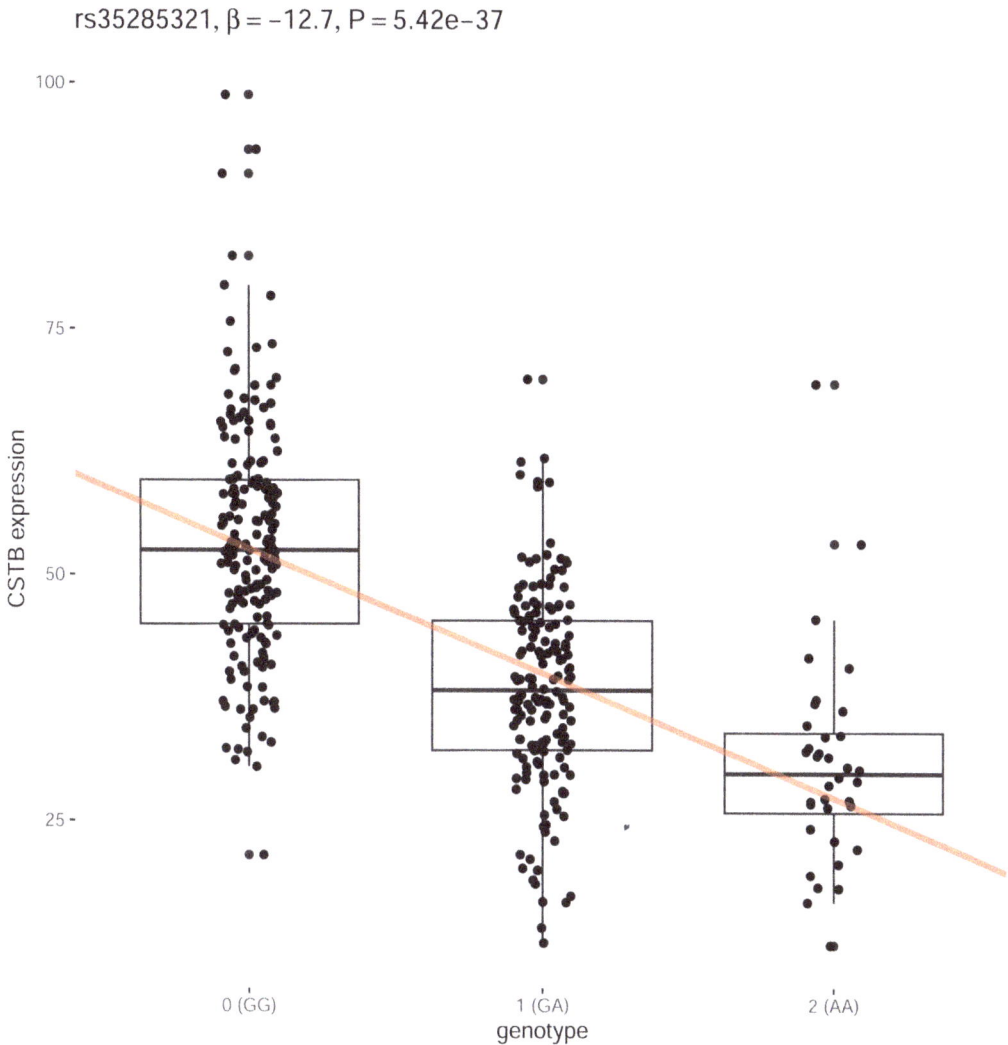

Figure 6.4 The expression of the gene *CTSB* in lymphobastoid cell lines derived from 373 subjects of European ancestry (data produced by the GEUVADIS project). Regression on the genotype of variant rs35285321, located ∼5 Kb upstream of the transcription start site, shows that the variant is indeed an eQTL for this gene. A so-called *jitter plot* is superimposed to the boxplot, showing the expression values of the individual subjects. These are randomly dispersed along a small part of the x axis so as to make them visible, but the actual value of the genotype is, of course, always exactly 0, 1, or 2.

example, if the SNP changed the binding site of a TF: Therefore, we can look at ChIP-seq data to see which TFs have binding sites overlapping this SNP, as shown in Figure 6.6.

Many TFs indeed have binding sites overlapping our SNP, confirming the active regulatory role of the region suggested by the chromHMM results. Moreover, a few

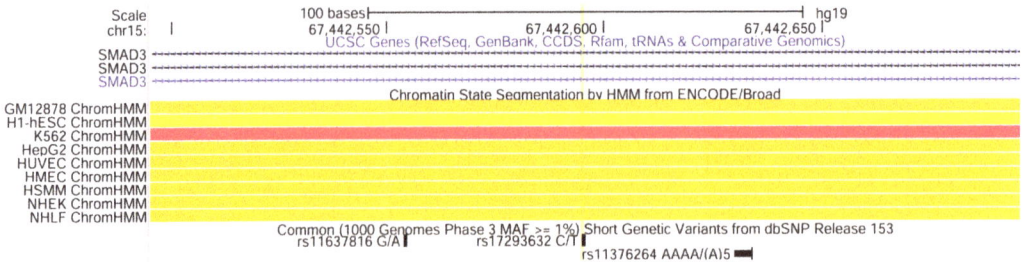

Figure 6.5 Genomic and epigenomic context of variant rs17293632 seen in the UCSC genome browser. The variant is located in an intron of gene *SMAD3*, in a region that chromHMM classifies as regulatory (red = promoter; yellow = enhancer) in all cell lines analyzed.

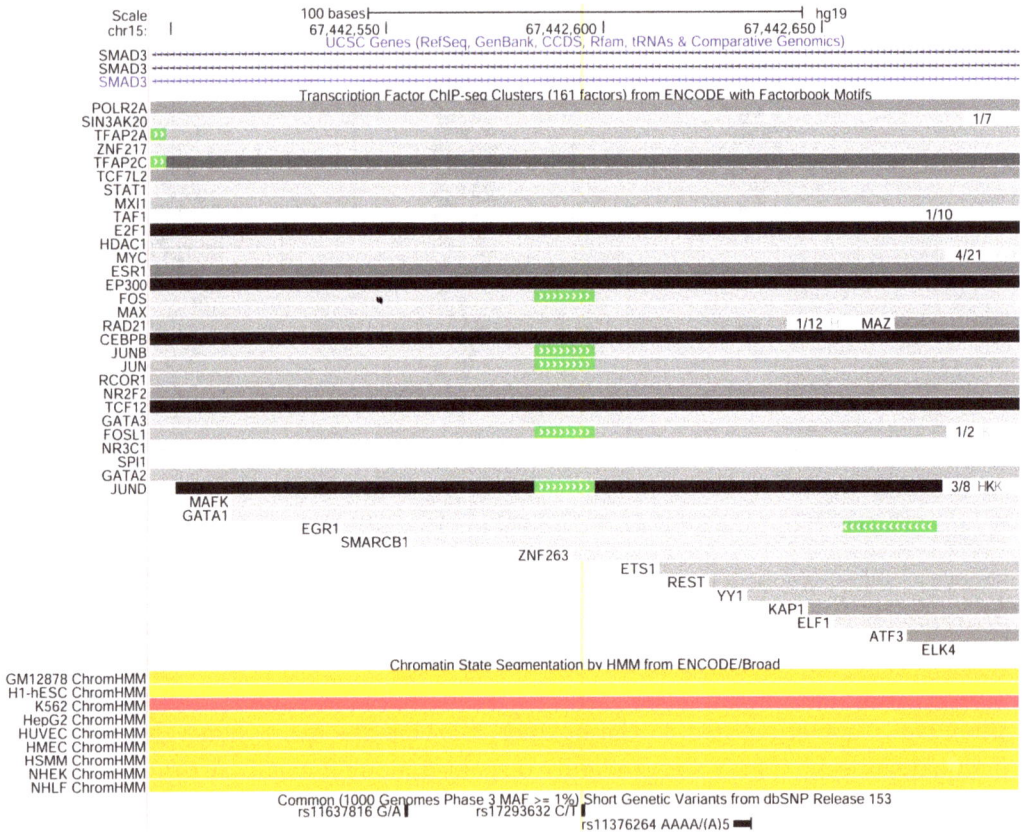

Figure 6.6 Binding sites derived from ChIP-seq experiments overlapping the variant of interest rs17293632. The track shown summarizes the ChIP-seq peaks (gray bars) of 161 TFs in 91 cell lines, and shows (in green) high-scoring motifs for the corresponding TFs.

Figure 6.7 The positional frequency matrix logo of the motif bound by transcription factor AP-1. The variant rs17293632 changes the tall C into a T, and thus is likely to significantly disrupt TF binding.

TFs (FOS, JUNB, JUN, FOSL1, and JUND) have a high-scoring motif overlapping the SNP, which could in principle be altered by the SNP. Actually, these TFs are all alternative components of the TF complex AP-1, which binds DNA with sequence specificity described by the motif shown in Figure 6.7. Our SNP changes the tall C into a T. As we saw in Chapter 5, a tall symbol implies that a large majority of the binding sites used to build the PFM had a C in that location, and thus that changing this C into a T might strongly affect TF binding.

Another piece of evidence supporting the functional relevance of the base affected by this variant comes from comparative genomics: The UCSC track displaying the multiple alignment of this region with 46 vertebrates (Figure 6.8) shows that the bases that compose the most informative part of the AP-1 motif, and in particular the one affected by our variant, are deeply conserved in placental mammals. As discussed in Chapter 1, conservation indicates negative selective pressure, and thus the functional relevance of this motif and of the specific C that is altered by the variant.

Finally, our interpretation of the mechanism connecting this variant to the phenotype requires the variant to cause a difference in gene expression between the carriers of the two alleles, and therefore to be an eQTL. We can check whether this is the case using the GTEx database [14], which contains eQTL data for 54 human tissues. Indeed rs17293632 is reported in GTEx to be a significant eQTL for *SMAD3* in two tissues (esophagus-mucosa and thyroid). This suggests that this variant is indeed able to change the expression of *SMAD3*, but two caveats should be noted. First, the effect of the variant turns out to be discordant in the two tissues (the alternate allele increases expression in the esophagus-mucosa but decreases it in the thyroid), which makes it difficult to generate mechanistic hypotheses. Second, GTEx reports the variant to be a significant eQTL also for two other nearby genes (*AAGAB* in whole blood and esophagus-mucosa, and *PIAS1* in thyroid). Therefore, all these considerations make it quite plausible that our variant alters disease susceptibility by altering gene expression, but are far from sufficient to postulate a precise mechanism, which requires ad hoc experiments in suitable cell lines, organoids, or animal models.

6.4.4 Predicting the Effect of Variants

eQTLs tell us which variants are associated to changes in gene expression, based on population-scale transcriptomic assays. A complementary approach for variant

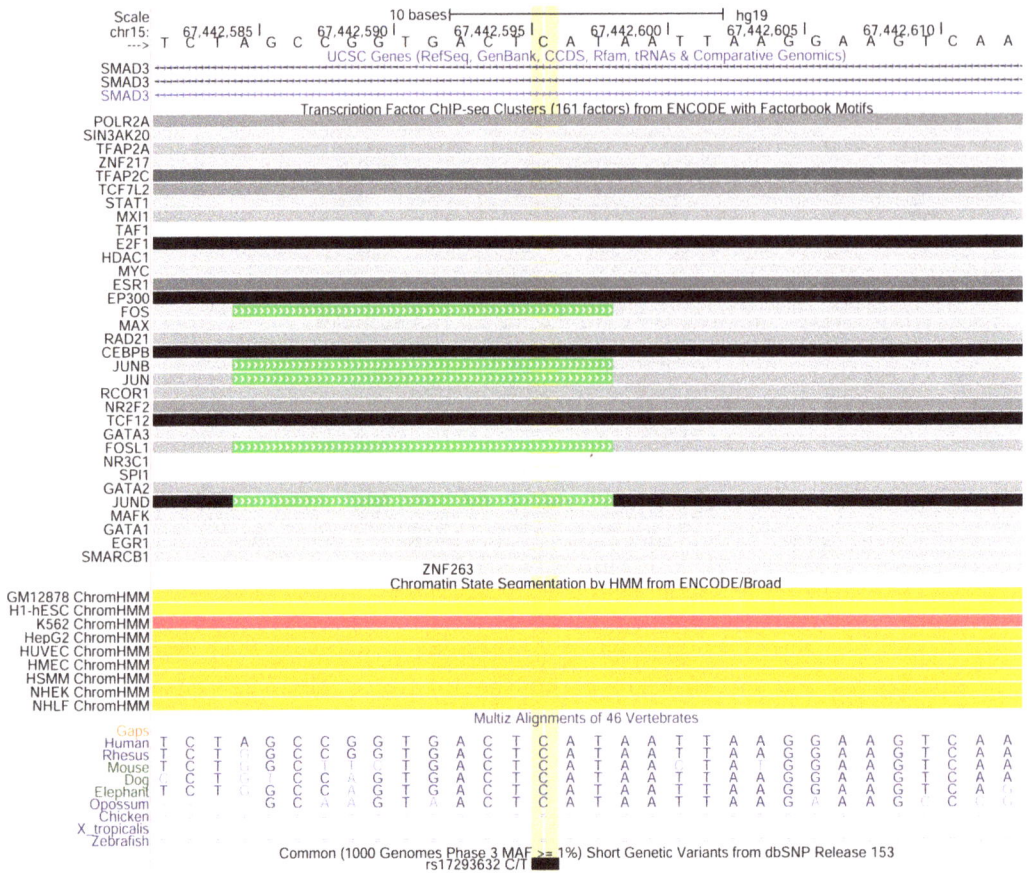

Figure 6.8 The "conservation" track of the UCSC genome is derived from the multiple alignment of the human genome with that of 46 vertebrates (only a few representative species are shown in this visualization). In particular it shows that the C nucleotide which gets changed to T by variant rs17293632 is highly conserved in placental mammals, suggesting negative selection and hence functional relevance.

interpretation and prioritization tries to *predict* which variants will affect some relevant epigenetic DNA state, such as the binding of a transcription factor or the state of the chromatin. Clearly, a variant affecting the epigenetic state is in turn likely to affect gene regulation, and thus in turn the trait of interest.

To be concrete, let us use TF binding as the state that we want to predict. We have already encountered the same classification problem in Chapter 5, where we used the PWM score as a classifier. Here we will not assume knowledge of a PWM and will describe a procedure that can be used also for other epigenetic states, such as chromatin accessibility, for which a PWM-based model is not suitable. The procedure goes as follows:

(a) use ChIP-seq data for the TF of interest and define the peaks as *positive sequences*, that is, sequences known to bind the TF;

(b) generate a comparable set of *negative sequences*, for example, by randomly selecting from the genome a set of sequences with the same numerosity and length distribution as the positive sequences. These are sequences for which we do not have any evidence of binding of the TF[13]. Ideally the negative sequences should match the positive ones in terms of general features that are likely to affect TF binding, such as GC content;

(c) define the *sequence features* that are most likely to be relevant in determining whether a sequence is positive or negative. Since we know that TFs recognize and bind short DNA sequences, a natural choice of features is given by the *k-mer* composition of the sequence: We extract from the sequence all its subsequences of length k (k-mers) and count how many times each possible k-mer appears. k is chosen to match the typical length of a TFBS, and sensible values are between $k = 3$ and $k = 10$. Remembering that TFs can recognize their binding site on both strands (see Chapter 5), we consider each k-mer and its reverse complement together. Thus, for example, choosing $k = 3$ the sequence

GATTATTAAT

gives the following k-mer counts:

k-mer	count
AAT/ATT	3
ATA/TAT	1
ATC/GAT	1
TAA/TTA	3

(d) use positive and negative sequences to train a model able to predict, from the k-mer composition of a sequence, whether it is positive or negative (i.e., whether it will bind or not the TF).

Thus we need a classifier:

> **i Classifier**
>
> A *classifier* is an algorithm able to classify some items into one of two or more classes (*labels*) based on a set of attributes of the items (*features*).

In our case the items are the sequences, labeled as positive or negative, and the features of a sequence are its k-mer counts. Many *machine learning* tools have

---

[13]In principle, one should check that no random sequence overlaps a positive one, but for most TFs the peaks cover such a small fraction of the genome that the random sequences are very unlikely to overlap a sizable number of positive ones.

been developed in the last ~50 years to approach such problems, including logistic regression, mentioned above in the context of GWAS, and, among others, *support vector machines*, *tree-based classifiers*, and *neural networks*. Here we will introduce support vector machines (SVMs), as they are relatively simple to understand, at least conceptually; allow us to define terms and discuss issues that are common to all machine learning classifiers; and finally happen to excel at the biological problem at hand, namely regulatory sequence classification.

6.4.5 Support Vector Machines

Before delving directly into the problem of classifying regulatory sequences, we will introduce SVMs in a much simpler and easily visualized context, in which the features used for prediction live in a two-dimensional space[14].

Suppose we are given the data points shown in Figure 6.9A. Each item is associated to two numerical features (x and y), and is represented as a point in the Cartesian plane. Moreover, each item is classified as positive (blue) or negative (red). Our goal is to find a way to predict whether a data point is positive or negative based on its coordinates x and y.

The simplest option is to find a straight line in the feature plane which separates as well as possible positives and negatives. For example, the straight line in Figure 6.9B does not do a good job, while the one in Figure 6.9C is much better (the precise way in which we evaluate the separation performance is described below). This classifier is called a *linear decision boundary*. A better classifier can be obtained by allowing the decision boundary not to be a straight line: This can be achieved by using a non-linear transformation on the coordinates of the original features and then seeking the best linear decision boundary in the transformed space, which will not be linear in the original space[15]. In this way, we can obtain the classification shown in Figure 6.9D, with further improvement with respect to Figure 6.9C.

The data used to determine the best decision boundary are called the *training data*: For these data we already know the label, so predicting it is not directly useful. However, if we now are given a new data point for which we know the value of the two features, but not the label, we can use the results of our training to predict the label based on which side of the decision boundary the new data point is located.

We still have to describe how exactly the performance of the SVM is evaluated, that is, in which precise sense the predictor shown in Figure 6.9D is better than the one in Figure 6.9C, which is better than the one in Figure 6.9B. A simple way to do this would be to count the number of misclassifications: Ten data points are misclassified in 6.9C, and only four in 6.9D, thus the latter SVM has better performance. Actually, SVMs use a more sophisticated criterion, which can be understood by considering a linear SVM achieving perfect separation (Figure 6.10). Here, the

---

[14]As opposed to k-mers, which live in a high-dimensional space. For example, for $k = 3$ each sequence is characterized by the counts of $4^3/2 = 32$ possible k-mers (we divide by 2 to account for the identification of a k-mer and its reverse complement). Thus, a sequence can be thought as being represented by a point in a 32-dimensional *feature space*.

[15]In practice, this is done by computing distances in the original space using a non-linear *kernel* instead of the Euclidean distance.

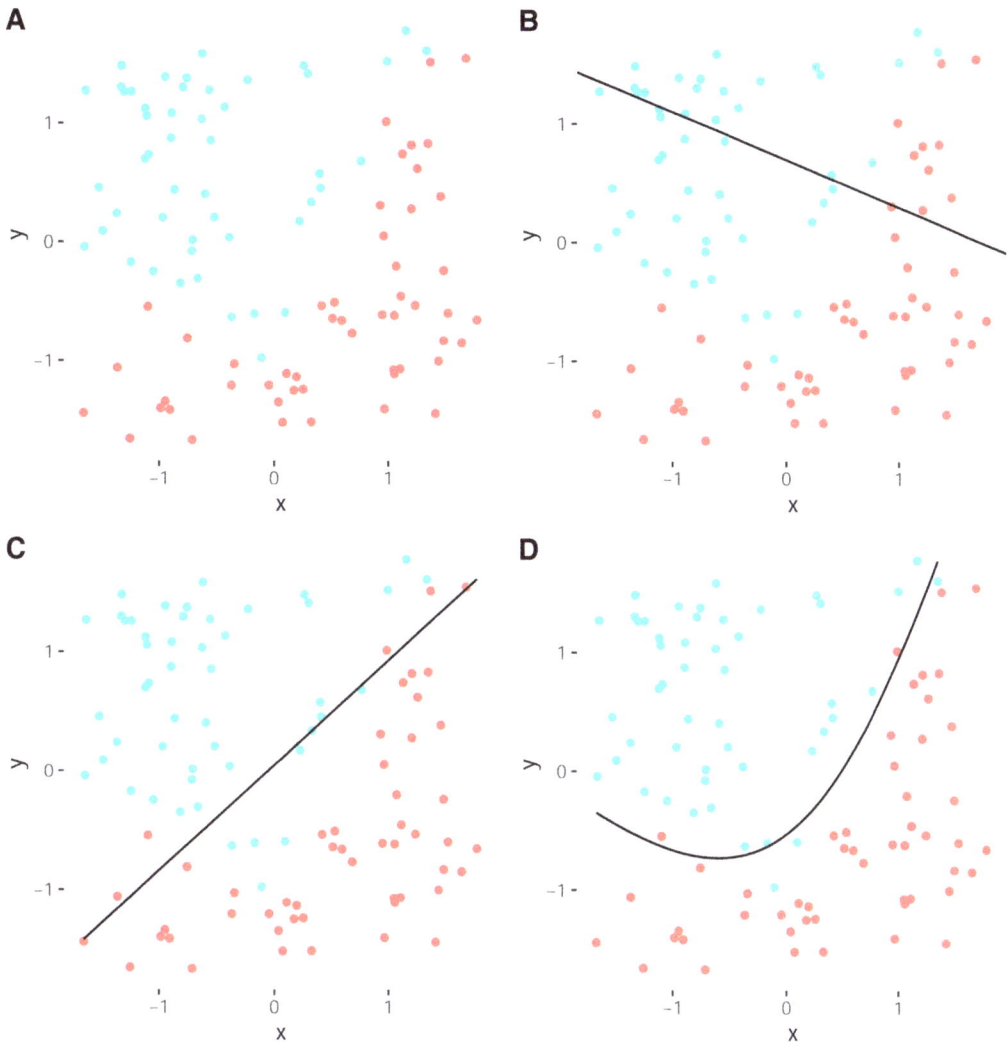

Figure 6.9 A: Data points labeled as *positive* (blue) or *negative* (red) are shown in a two-dimensional feature space. We want to predict the label based on the value of the features. B: A straight line in feature space that does not separate well positive from negative points. C: Linear SVM finds the straight line that best separates positives and negatives. D: Better separation can be obtained using a non-linear SVM.

linear SVM identifies the solid line as the best decision boundary. However, also the dashed line achieves perfect separation of positive and negative points.

The reason why the solid line is better than the dotted one is that it has a higher *margin*, defined as the distance between the line and the closest data point. Intuitively, new data points on which we want to predict the labels will be better predicted using the solid than the dashed line: A new, negative data point is more likely to fall above the decision boundary, and thus to be misclassified, if we use the dashed line than if we use the solid one. Therefore, when, as in this case, it is possible to achieve

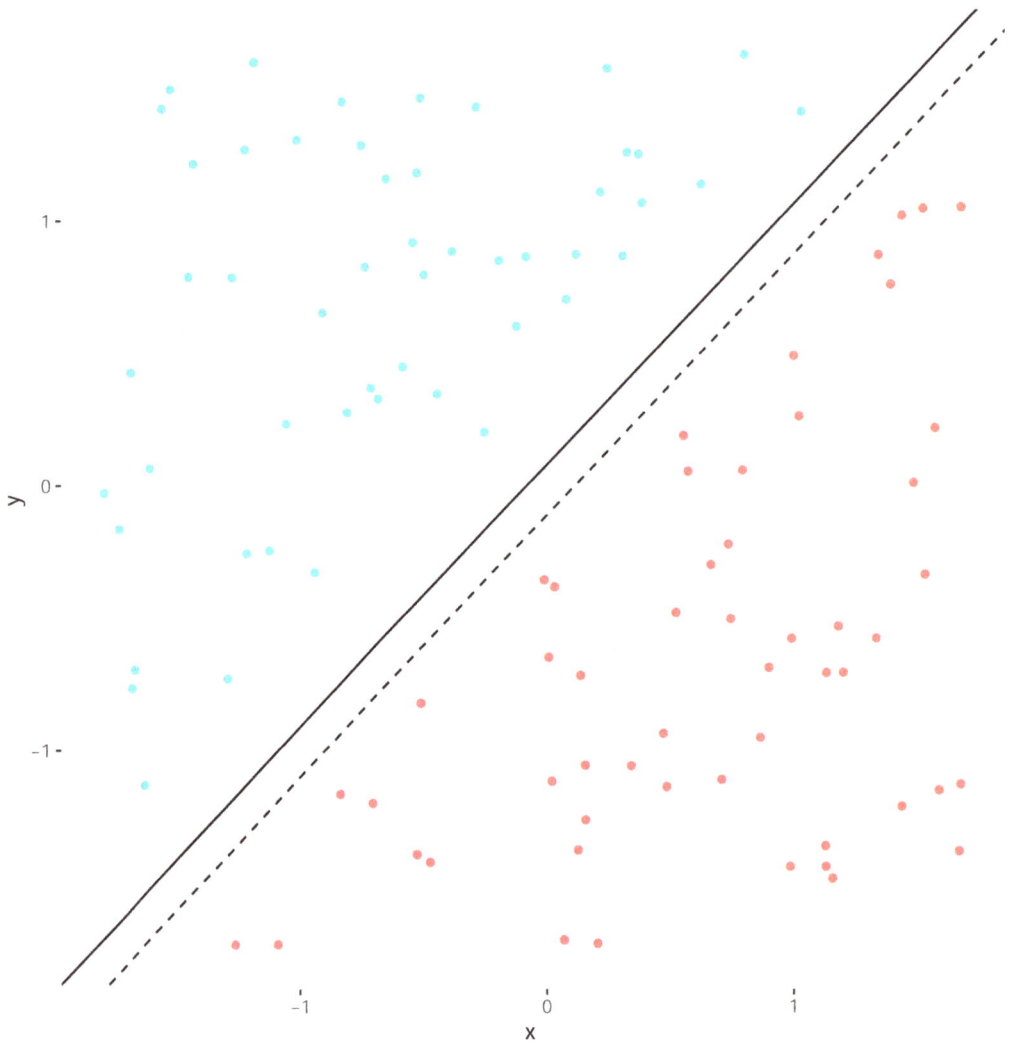

Figure 6.10 In this case a linear SVM achieves perfect separation between the two classes. The best decision boundary is the solid line. It is a better decision boundary than the dashed line (which would also achieve perfect separation) because of the larger margin, which decreases the risk of misclassification of future data points.

perfect separation, the best decision boundary is chosen as the one that maximizes the margin; in most cases perfect separation cannot be achieved, and SVMs maximize a suitable combination of accuracy (number of data points accurately classified) and margin (now defined as the distance from the closest *correctly classified* data point).

6.4.6 Training and Cross-Validation

The final goal of a classifier is to predict the class of new data for which we know the values of the features but not that of the label. On the other hand, the *training* of the classifier must be done on data for which we know both features and labels. This leads to the potential problem of *overfitting*:

> **i Overfitting**
>
> A classifier is said to be *overfitting* if it learns and uses for prediction characteristics of the training data that are not shared by the data sets on which it is used for actual prediction.

Thus overfitting increases the performance of the classifier on the training data, but decreases its *generalization*, that is, its performance on data not used for training[16]. For example, by using a more complex non-linear transformation[17] one can achieve perfect separation for the data of Figure 6.9, as shown in Figure 6.11. A sufficiently complex non-linear transformation can achieve perfect separation, but the generalization properties of this decision boundary (i.e. its performance on data not used for training) might be in doubt if it reflects peculiarities of the specific data used for training.

One way to control overfitting is called *cross-validation*, and consists in randomly subdividing the data for which we know both features and labels into two subsets: The first subset, the *training set*, is used to train the model, hence in our case to find the best-performing SVM. The performance of this SVM is then assessed on the second subset, the *validation set*, for example by computing its sensitivity and specificity, or by building a ROC curve and computing the AUC (see Chapter 5)[18]. Since the data in the validation set were not used in the training, the accidental features of the former will not be present in the latter: Thus, overfitting will increase the performance of the model on the training set, but decrease it on the validation set.

If we just divide the data into one training and one validation set, we get a single evaluation of the classifier performance. To obtain a more robust result, and an assessment of the variability of the performance, a more sophisticated method called *k-fold* cross-validation is often used. Suppose we have a total of N data points for which we know features and label, thus potentially available for training. We then

(a) partition the N data points into k subsets of (approximately) equal size. These k subsets are called the *folds*;

(b) use each fold as the validation set of a model trained on the union of all the other folds (thus, each model is trained on $N \cdot \frac{k-1}{k}$ data points).

---

[16]The following brief discussion of overfitting and how to control it is not specific of SVMs, but applies to all machine-learning methods of classification.

[17]Specifically, by using a polynomial kernel of degree 6 rather than the one of degree 2 used in Figure 6.9D.

[18]This can be done because SVMs provide, for each predicted point, not just the predicted label but a continuous value measuring the signed distance from the decision boundary. By varying the cutoff on this value one can build a ROC curve and compute its AUC.

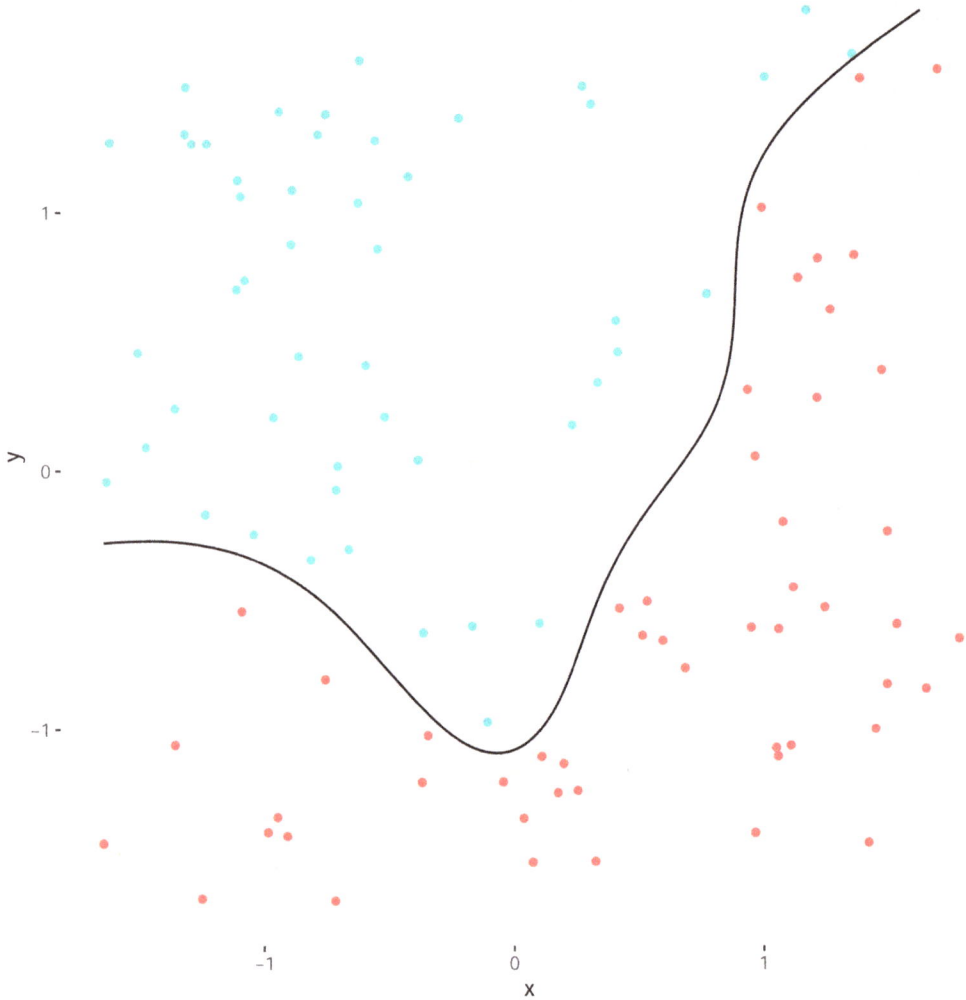

Figure 6.11 By using a complex non-linear transformation of the coordinates we can achieve perfect separation in the example of Figure 6.9. However overfitting might have taken place, in which accidental characteristics of the specific data set used for training get learned by the model, compromising its ability to predict the labels for new data.

In this way, we obtain k assessment of the performance of our classifier. Increasing values of k lead to a higher number of evaluations of the performance conducted on smaller validation sets, so that each individual evaluation is noisier. Popular choices are $k = 5$ or $k = 10$. Leave-one-out cross-validation, that we saw above in the context of the prediction of causal genes for Mendelian diseases, can be considered as the extreme case $k = N$, as the model is successively trained on all data points except for one, and the performance is evaluated on the excluded point.

6.4.7 A SVM to Predict Chromatin Features

In particular, SVMs using k-mer counts as features can be developed to predict epigenetic features (such as TF binding, chromatin accessibility, etc.). We will briefly discuss the model introduced in [25] and successively improved in [13], and in particular how the model can be used to predict the effects of genetic variants on the epigenetic feature of interest [24].

The original paper [25] focused on predicting, from the DNA sequence, the enhancer status of a genomic region in a given cellular context. Positive sequences were derived from the ChIP-seq peaks of EP300, a well-known transcriptional coactivator whose binding is strongly associated to enhancer activity, and suitably matched random genomic sequences were used as negatives. Then a SVM was trained with k-mer counts as features and a non-linear kernel, and its performance was assessed by showing, in particular, that a SVM trained on mouse enhancers could actually predict also human enhancers.

The SVM was then applied to the whole mouse genome to identify potential new enhancers, that is, sequences predicted by the SVM to be enhancers but not bound by EP300 according to the ChIP-seq data. These sequences could in principle be (a) false positives, that is, sequences that are not truly enhancers but are erroneously predicted to be enhancers by the SVM, or (b) true enhancers that were not discovered using ChIP-seq, possibly because of a too stringent peak detection strategy[19]. Using independent data sources, the authors showed that a large fraction of these sequences fall into case (b). Therefore, an SVM based on k-mer counts can produce new biological information by predicting enhancers that are not readily identifiable from ChIP-seq assays.

Besides being useful for predicting new enhancers, this SVM generates information about which k-mers are more strongly associated to enhancer status. In fact, the optimal decision boundary found by the SVM is expressed by assigning to each feature, that is, to each k-mer, a *weight*: A positive (negative) weight means that the presence of the k-mer is predictive of a positive (negative) sequence. This allows us to use the SVM not only for prediction, but also to gain biological insight by finding the k-mers that are most predictive of enhancer status. For example, when the SVM was trained on embryonic mouse forebrain tissue, the highest-weight k-mers could be recognized as binding sites of TFs involved in embryonic nervous system development.

Moreover, the k-mer weights allow us to predict which variants will significantly affect the epigenetic state predicted by the SVM. As an example, we will turn to a dataset explored in [13][20]. Here, the epigenetic feature of interest is the binding of CTCF, and the training data are corresponding ChIP-seq assays.

---

[19]For a more nuanced discussion see the original paper.

[20]This paper presents an improved version of the SVM using *gapped k-mers* as features, which leads to a considerable improvement in performance. As we will discuss only the k-mer weights and their use in the prediction of variant effects, we will not describe this improved SVM, and refer the reader to the original publication.

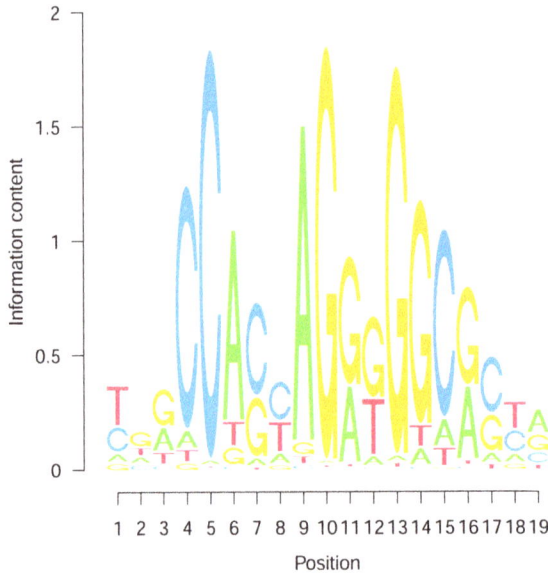

Figure 6.12 The logo of the CTCF PFM shows that the top k-mers found by the SVM are indeed good matches, as expected.

Using $k = 10$, the top k-mers by weight are shown in the table below:

k-mer	weight
CCACTAGGGG	6.859
CCACCAGGGG	6.695
CCACTAGAGG	6.46
CACCTAGTGG	6.187
CACCTGGTGG	6.133
CACCAGGGGG	5.907

These are the k-mers with the highest *positive* weight, so they should be good matches with the CTCF PWM, a fact that is immediately apparent by looking at the logo in Figure 6.12.

6.4.8 Variant Effect Prediction with the SVM

These weights naturally suggest a way to predict the effect of a genetic variant on the epigenetic feature of interest, in this case CTCF binding: A variant will be predicted to have a strong effect if it changes a k-mer of high weight into one of low weight, or vice versa. Consider, for example, the SNP rs16946083, located on chromosome 15, where the reference sequence carries a C, while the alternate allele is G. The alternate allele has a frequency of ~9%, making this a common SNP. To predict the effect of this variant on the binding of CTCF, we have to consider all the k-mers in the genomic sequence that are changed by the variant. If $k = 10$, we need to consider

10 k-mers derived from the genomic sequence of length 19 centered at the position of the variant. The reference sequence is:

GTTTCACTA<u>C</u>AGGGCGCTG

and the alternate is:

GTTTCACTA<u>G</u>AGGGCGCTG

To evaluate the effect of the variant, we compute the total weight of the 10 k-mers found in the alternate sequence minus the same for the reference sequence: This quantity is called *deltaSVM*. A positive (negative) result suggests that the alternate sequence is more (less) likely to bind CTCF compared with the reference sequence. In our case, we get:

k-mer (ref)	weight	k-mer (alt)	weight
GTTTCACTAC	0.2435	GTTTCACTAG	0.09264
TTTCACTACA	0.1707	TTTCACTAGA	0.194
TTCACTACAG	0.222	TTCACTAGAG	0.3393
TCACTACAGG	0.3445	TCACTAGAGG	1.516
CACTACAGGG	1.401	CACTAGAGGG	5.756
ACTACAGGGC	1.043	ACTAGAGGGC	4.403
CTACAGGGCG	1.018	CTAGAGGGCG	3.903
TACAGGGCGC	0.9303	TAGAGGGCGC	3.845
ACAGGGCGCT	0.7173	AGAGGGCGCT	4.053
CAGGGCGCTG	0.2737	GAGGGCGCTG	1.872
total	6.365		25.97

so that the deltaSVM is ∼19.6 and we predict the alternate allele to favor the binding of CTCF (as it could also be guessed by comparing the two sequences to the PFM). The value of the deltaSVM for a variant of interest is then compared with the deltaSVM values for suitably chosen control SNPs (e.g., randomly chosen SNPs, or candidate causal SNPs in the same locus) to prioritize the variants most likely to change the epigenetic state of the region, and thus gene expression and eventually the trait of interest.

FURTHER READING

Aerts, S., Lambrechts, D., Maity, S., Van Loo, P., Coessens, B., De Smet, F., Tranchevent, L., De Moor, B., Marynen, P., Hassan, B., Carmeliet, P. & Moreau, Y. Gene prioritization through genomic data fusion. *Nat Biotechnol.* **24**, 537–544 (2006).

Aguet, F., Brown, A., Castel, S., Davis, J., He, Y., Jo, B., Mohammadi, P., Park, Y., Parsana, P., Segrè, A., et al. Genetic effects on gene expression across human tissues. *Nature.* **550**, 204–213 (2017).

Lappalainen, T., Sammeth, M., Friedländer, M., Hoen, P., Monlong, J., Rivas, M., Gonzàlez-Porta, M., Kurbatova, N., Griebel, T., Ferreira, P., et al. Transcriptome and genome sequencing uncovers functional variation in humans. *Nature.* **501**, 506–511 (2013).

Lee, D., Gorkin, D., Baker, M., Strober, B., Asoni, A., McCallion, A. & Beer, M. A method to predict the impact of regulatory variants from DNA sequence. *Nat Genet.* **47**, 955–961 (2015).

Lee, D., Karchin, R. & Beer, M. Discriminative prediction of mammalian enhancers from DNA sequence. *Genome Res.* **21**, 2167–2180 (2011).

Tarpey, P., Smith, R., Pleasance, E., Whibley, A., Edkins, S., Hardy, C., O'Meara, S., Latimer, C., Dicks, E., Menzies, et al. A systematic, large-scale resequencing screen of X-chromosome coding exons in mental retardation. *Nat Genet.* **41**, 535–543 (2009).

Software Package Used

The data were analyzed and the figures produced using the UCSC Genome Browser, the Integrative Genome Viewer, and the following R/Bioconductor libraries:

ape, aplot, Biostrings, BSgenome.Hsapiens.UCSC.hg19, BSgenome.Hsapiens. UCSC.hg38.masked, cccd, clusterProfiler, cowplot, data.table, dendextend, dplyr, e1071, edgeR, GenomicRanges, GEOquery, ggfortify, ggplot2, ggpubr, ggtree, gkmSVM, gridExtra, igraph, JASPAR2020, kernlab, knitr, limma, markovchain, miRBaseConverter, monocle, msa, org.Hs.eg.db, org.Sc.sgd.db, pander, patchwork, phangorn, pheatmap, plotrix, plyr, pwalign, qqman, readxl, regioneR, reshape2, rhdf5, ROCR, Rtsne, scater, scran, scRNAseq, scuttle, seqinr, Seurat, survival, survminer, TFBSTools, TxDb.Hsapiens.UCSC.hg19.knownGene, TxDb.Hsapiens.UCSC.hg38. knownGene, vcfR, velociraptor, zebrafish.db and zip

Bibliography

[1] S. F. Altschul, W. Gish, W. Miller, E. W. Myers, and D. J. Lipman. Basic local alignment search tool. *J Mol Biol*, 215(3):403–410, October 1990.

[2] Simon Anders and Wolfgang Huber. Differential expression analysis for sequence count data. *Genome Biol*, 11(10):R106, October 2010.

[3] Ayca Arslan-Ergul and Michelle M Adams. Gene expression changes in aging zebrafish (Danio rerio) brains are sexually dimorphic. *BMC Neurosci*, 15:29, February 2014.

[4] Nicolas L. Bray, Harold Pimentel, Páll Melsted, and Lior Pachter. Near-optimal probabilistic RNA-seq quantification. *Nat Biotechnol*, 34(5):525–527, May 2016.

[5] Nipun Chopra, Ruizhi Wang, Bryan Maloney, Kwangsik Nho, John S. Beck, Naemeh Pourshafie, Alexander Niculescu, Andrew J. Saykin, Carlo Rinaldi, Scott E. Counts, and Debomoy K. Lahiri. MicroRNA-298 reduces levels of human amyloid-β precursor protein (APP), β-site APP-converting enzyme 1 (BACE1) and specific tau protein moieties. *Mol Psychiatry*, 26(10):5636–5657, October 2021.

[6] Luis A. Corchete, Elizabeta A. Rojas, Diego Alonso-López, Javier De Las Rivas, Norma C. Gutiérrez, and Francisco J. Burguillo. Systematic comparison and assessment of RNA-seq procedures for gene expression quantitative analysis. *Sci Rep*, 10(1):19737, November 2020. Number: 1 Publisher: Nature Publishing Group.

[7] J L DeRisi, V R Iyer, and P O Brown. Exploring the metabolic and genetic control of gene expression on a genomic scale. *Science*, 278(5338):680–686, October 1997.

[8] Richard Durbin, Sean R Eddy, Anders Krogh, and Graeme Mitchison. *Biological Sequence Analysis*. Cambridge University press, 1998.

[9] Yaniv Erlich, Simon Edvardson, Emily Hodges, Shamir Zenvirt, Pramod Thekkat, Avraham Shaag, Talya Dor, Gregory J. Hannon, and Orly Elpeleg. Exome sequencing and disease-network analysis of a single family implicate a mutation in KIF1A in hereditary spastic paraparesis. *Genome Res*, 21(5):658–664, May 2011.

[10] Jason Ernst and Manolis Kellis. Discovery and characterization of chromatin states for systematic annotation of the human genome. *Nat Biotechnol*, 28(8):817–825, August 2010.

[11] Jason Ernst, Pouya Kheradpour, Tarjei S Mikkelsen, Noam Shoresh, Lucas D Ward, Charles B Epstein, Xiaolan Zhang, Li Wang, Robbyn Issner, Michael Coyne, Manching Ku, Timothy Durham, Manolis Kellis, and Bradley E Bernstein. Mapping and analysis of chromatin state dynamics in nine human cell types. *Nature*, 473(7345):43–49, May 2011.

[12] Kyle Kai-How Farh, Alexander Marson, Jiang Zhu, Markus Kleinewietfeld, William J Housley, Samantha Beik, Noam Shoresh, Holly Whitton, Russell J H Ryan, Alexander A Shishkin, Meital Hatan, Marlene J Carrasco-Alfonso, Dita Mayer, C John Luckey, Nikolaos A Patsopoulos, Philip L De Jager, Vijay K Kuchroo, Charles B Epstein, Mark J Daly, David A Hafler, and Bradley E Bernstein. Genetic and epigenetic fine mapping of causal autoimmune disease variants. *Nature*, 518(7539):337–343, February 2015.

[13] Mahmoud Ghandi, Dongwon Lee, Morteza Mohammad-Noori, and Michael A. Beer. Enhanced regulatory sequence prediction using gapped k-mer features. *PLoS Comput Biol*, 10(7):e1003711, July 2014.

[14] GTEx Consortium. The GTEx Consortium atlas of genetic regulatory effects across human tissues. *Science*, 369(6509):1318–1330, September 2020.

[15] Gunsagar S. Gulati, Shaheen S. Sikandar, Daniel J. Wesche, Anoop Manjunath, Anjan Bharadwaj, Mark J. Berger, Francisco Ilagan, Angera H. Kuo, Robert W. Hsieh, Shang Cai, Maider Zabala, Ferenc A. Scheeren, Neethan A. Lobo, Dalong Qian, Feiqiao B. Yu, Frederick M. Dirbas, Michael F. Clarke, and Aaron M. Newman. Single-cell transcriptional diversity is a hallmark of developmental potential. *Science*, 367(6476):405–411, January 2020.

[16] Yuhan Hao, Stephanie Hao, Erica Andersen-Nissen, William M. Mauck, Shiwei Zheng, Andrew Butler, Maddie J. Lee, Aaron J. Wilk, Charlotte Darby, Michael Zager, Paul Hoffman, Marlon Stoeckius, Efthymia Papalexi, Eleni P. Mimitou, Jaison Jain, Avi Srivastava, Tim Stuart, Lamar M. Fleming, Bertrand Yeung, Angela J. Rogers, Juliana M. McElrath, Catherine A. Blish, Raphael Gottardo, Peter Smibert, and Rahul Satija. Integrated analysis of multimodal single-cell data. *Cell*, 184(13):3573–3587.e29, June 2021.

[17] Brian P. Hermann, Keren Cheng, Anukriti Singh, Lorena Roa-De La Cruz, Kazadi N. Mutoji, I.-Chung Chen, Heidi Gildersleeve, Jake D. Lehle, Max Mayo, Birgit Westernströer, Nathan C. Law, Melissa J. Oatley, Ellen K. Velte, Bryan A. Niedenberger, Danielle Fritze, Sherman Silber, Christopher B. Geyer, Jon M. Oatley, and John R. McCarrey. The Mammalian Spermatogenesis Single-Cell Transcriptome, from Spermatogonial Stem Cells to Spermatids. *Cell Rep*, 25(6):1650–1667.e8, November 2018.

[18] Jaime Huerta-Cepas, Hernán Dopazo, Joaquín Dopazo, and Toni Gabaldón. The human phylome. *Genome Biol*, 8(6):R109, 2007.

[19] Pablo A. Jaskowiak, Ricardo J. G. B. Campello, and Ivan G. Costa. On the selection of appropriate distances for gene expression data clustering. *BMC Bioinformatics*, 15 Suppl 2:S2, 2014.

[20] Lukas F. K. Kuderna, Hong Gao, Mareike C. Janiak, Martin Kuhlwilm, Joseph D. Orkin, Thomas Bataillon, et al. A global catalog of whole-genome diversity from 233 primate species. Science, 380(6648):906, 2023.

[21] Gioele La Manno, Ruslan Soldatov, Amit Zeisel, Emelie Braun, Hannah Hochgerner, Viktor Petukhov, Katja Lidschreiber, Maria E. Kastriti, Peter Lönnerberg, Alessandro Furlan, Jean Fan, Lars E. Borm, Zehua Liu, David van Bruggen, Jimin Guo, Xiaoling He, Roger Barker, Erik Sundström, Gonçalo Castelo-Branco, Patrick Cramer, Igor Adameyko, Sten Linnarsson, and Peter V. Kharchenko. RNA velocity of single cells. *Nature*, 560(7719):494–498, August 2018.

[22] Tuuli Lappalainen, Michael Sammeth, Marc R. Friedländer, Peter A. C. 't Hoen, Jean Monlong, Manuel A. Rivas, Mar Gonzàlez-Porta, Natalja Kurbatova, Thasso Griebel, Pedro G. Ferreira, Matthias Barann, Thomas Wieland, Liliana Greger, Maarten van Iterson, Jonas Almlöf, Paolo Ribeca, Irina Pulyakhina, Daniela Esser, Thomas Giger, Andrew Tikhonov, Marc Sultan, Gabrielle Bertier, Daniel G. MacArthur, Monkol Lek, Esther Lizano, Henk P. J. Buermans, Ismael Padioleau, Thomas Schwarzmayr, Olof Karlberg, Halit Ongen, Helena Kilpinen, Sergi Beltran, Marta Gut, Katja Kahlem, Vyacheslav Amstislavskiy, Oliver Stegle, Matti Pirinen, Stephen B. Montgomery, Peter Donnelly, Mark I. McCarthy, Paul Flicek, Tim M. Strom, Geuvadis Consortium, Hans Lehrach, Stefan Schreiber, Ralf Sudbrak, Angel Carracedo, Stylianos E. Antonarakis, Robert Häsler, Ann-Christine Syvänen, Gert-Jan van Ommen, Alvis Brazma, Thomas Meitinger, Philip Rosenstiel, Roderic Guigó, Ivo G. Gut, Xavier Estivill, and Emmanouil T. Dermitzakis. Transcriptome and genome sequencing uncovers functional variation in humans. *Nature*, 501(7468):506–511, September 2013.

[23] Charity W. Law, Yunshun Chen, Wei Shi, and Gordon K. Smyth. voom: precision weights unlock linear model analysis tools for RNA-seq read counts. *Genome Biology*, 15(2):R29, February 2014.

[24] Dongwon Lee, David U. Gorkin, Maggie Baker, Benjamin J. Strober, Alessandro L. Asoni, Andrew S. McCallion, and Michael A. Beer. A method to predict the impact of regulatory variants from DNA sequence. *Nat Genet*, 47(8):955–961, August 2015.

[25] Dongwon Lee, Rachel Karchin, and Michael A. Beer. Discriminative prediction of mammalian enhancers from DNA sequence. *Genome Res*, 21(12):2167–2180, December 2011.

[26] Michael I. Love, Wolfgang Huber, and Simon Anders. Moderated estimation of fold change and dispersion for RNA-seq data with DESeq2. *Genome Biol*, 15(12):550, December 2014.

[27] John C. Marioni, Christopher E. Mason, Shrikant M. Mane, Matthew Stephens, and Yoav Gilad. RNA-seq: an assessment of technical reproducibility and comparison with gene expression arrays. *Genome Res*, 18(9):1509–1517, September 2008.

[28] Jasmine L. May, Fotini M. Kouri, Lisa A. Hurley, Juan Liu, Serena Tommasini-Ghelfi, Yanrong Ji, Peng Gao, Andrea E. Calvert, Andrew Lee, Navdeep S. Chandel, Ramana V. Davuluri, Craig M. Horbinski, Jason W. Locasale, and Alexander H. Stegh. IDH3α regulates one-carbon metabolism in glioblastoma. *Sci Adv*, 5(1):eaat0456, January 2019.

[29] Leland McInnes, John Healy, and James Melville. UMAP: Uniform Manifold Approximation and Projection for Dimension Reduction, September 2020. arXiv:1802.03426 [cs, stat].

[30] Vamsi K. Mootha, Cecilia M. Lindgren, Karl-Fredrik Eriksson, Aravind Subramanian, Smita Sihag, Joseph Lehar, Pere Puigserver, Emma Carlsson, Martin Ridderstråle, Esa Laurila, Nicholas Houstis, Mark J. Daly, Nick Patterson, Jill P. Mesirov, Todd R. Golub, Pablo Tamayo, Bruce Spiegelman, Eric S. Lander, Joel N. Hirschhorn, David Altshuler, and Leif C. Groop. PGC-1alpha-responsive genes involved in oxidative phosphorylation are coordinately downregulated in human diabetes. *Nat Genet*, 34(3):267–273, July 2003.

[31] Katherine S. Pollard, Melissa J. Hubisz, Kate R. Rosenbloom, and Adam Siepel. Detection of nonneutral substitution rates on mammalian phylogenies. *Genome Res*, 20(1):110–121, January 2010.

[32] Matthew E Ritchie, Belinda Phipson, Di Wu, Yifang Hu, Charity W Law, Wei Shi, and Gordon K Smyth. limma powers differential expression analyses for RNA-sequencing and microarray studies. *Nucleic Acids Res*, 43(7):e47, April 2015.

[33] Mark D. Robinson, Davis J. McCarthy, and Gordon K. Smyth. edgeR: a Bioconductor package for differential expression analysis of digital gene expression data. *Bioinformatics*, 26(1):139–140, January 2010.

[34] Adrian L. Sanborn, Suhas S. P. Rao, Su-Chen Huang, Neva C. Durand, Miriam H. Huntley, Andrew I. Jewett, Ivan D. Bochkov, Dharmaraj Chinnappan, Ashok Cutkosky, Jian Li, Kristopher P. Geeting, Andreas Gnirke, Alexandre Melnikov, Doug McKenna, Elena K. Stamenova, Eric S. Lander, and Erez Lieberman Aiden. Chromatin extrusion explains key features of loop and domain formation in wild-type and engineered genomes. *Proc Nat Acad Sci*, 112(47):E6456–E6465, November 2015.

[35] Charlotte Soneson and Mauro Delorenzi. A comparison of methods for differential expression analysis of RNA-seq data. *BMC Bioinform*, 14(1):91, March 2013.

[36] Aravind Subramanian, Pablo Tamayo, Vamsi K. Mootha, Sayan Mukherjee, Benjamin L. Ebert, Michael A. Gillette, Amanda Paulovich, Scott L. Pomeroy, Todd R. Golub, Eric S. Lander, and Jill P. Mesirov. Gene set enrichment analysis: A knowledge-based approach for interpreting genome-wide expression profiles. *Proc Natl Acad Sci USA*, 102(43):15545–15550, October 2005.

[37] S Tavazoie, J D Hughes, M J Campbell, R J Cho, and G M Church. Systematic determination of genetic network architecture. *Nat Genet*, 22(3):281–285, July 1999.

[38] Léon-Charles Tranchevent, Amin Ardeshirdavani, Sarah ElShal, Daniel Alcaide, Jan Aerts, Didier Auboeuf, and Yves Moreau. Candidate gene prioritization with Endeavour. *Nucleic Acids Res*, 44(W1):W117–121, July 2016.

[39] Cole Trapnell, Davide Cacchiarelli, Jonna Grimsby, Prapti Pokharel, Shuqiang Li, Michael Morse, Niall J. Lennon, Kenneth J. Livak, Tarjei S. Mikkelsen, and John L. Rinn. The dynamics and regulators of cell fate decisions are revealed by pseudotemporal ordering of single cells. *Nat Biotechnol*, 32(4):381–386, April 2014.

[40] Laurens van der Maaten and Geoffrey Hinton. Visualizing Data using t-SNE. *J Mach Learn Res*, 9:2579–2605, November 2008.

[41] Yong Zhang, Tao Liu, Clifford A Meyer, Jérôme Eeckhoute, David S Johnson, Bradley E Bernstein, Chad Nusbaum, Richard M Myers, Myles Brown, Wei Li, and X Shirley Liu. Model-based analysis of ChIP-Seq (MACS). *Genome Biol*, 9(9):R137, September 2008.

Index

For Product Safety Concerns and Information please contact our EU representative GPSR@taylorandfrancis.com
Taylor & Francis Verlag GmbH, Kaufingerstraße 24, 80331 München, Germany

www.ingramcontent.com/pod-product-compliance
Lightning Source LLC
Chambersburg PA
CBHW061419210326
41598CB00035B/6270